D0504764

Electrical machines for technicians and technician engineers

Electrical machines for technicians and technician engineers

Stefan F. Jurek,

B.Sc., C.Eng., M.I.E.E., M.I.Mech.E.

Head of Department of Engineering Technology,
Mid-Herts College,
Welwyn Garden City, Herts.

Longman

LONGMAN GROUP LIMITED
London

*Associated companies, branches and representatives
throughout the world*

© Longman Group Limited 1972

First published 1972

ISBN 0 582 42601·4

Printed in Great Britain by
William Clowes & Sons Limited
London, Colchester and Beccles

Contents

Preface

This book is written specifically for technicians and technician engineers and presents the theory of transformers and rotating machines to students following courses for the City & Guilds Technician Certificates, Ordinary National Certificates and Diplomas in Engineering and Ordinary National Diploma in Technology (Engineering). It should also prove useful as an introduction for H.N.C., H.N.D., and degree students, as well as a reference for technicians who have completed their studies.

The text emphasises the unity underlying all electrical machines in line with the modern generalised approach, and relates the theoretical considerations to actual machines by means of many diagrams and illustrations. In particular, the 'steady-state' theory of rotating machines is developed around the magnetic field in the air-gap through which electrical energy is transformed into mechanical energy and vice versa. This treatment stresses the reversibility of each machine as an electro-mechanical energy converter. It is felt that to attempt to write a textbook for a particular year of a given course is impracticable in view of the continuous and rapid revision of syllabuses; instead the present volume is so arranged that self-contained paragraphs may be chosen for an appropriate subject and year of a given technician course.

Acknowledgements

I wish to express my gratitude, first, to Professor M. G. Say whose lectures in the mid-'fifties on the Unified Approach to Electrical Machines were the starting point of my interest in this subject, to my colleagues Messrs R. Gill, D. Miles and R. Sherman who have read the manuscript and made helpful suggestions, to various 'expert readers' whose criticisms lead to many revisions and improvements of the book, and to all who have helped with the preparation of the manuscript and its diagrams.

My thanks are also due to the City & Guilds of London Institute for permission to use their past examination questions and to various manufacturers who so readily provided information and illustrations. All are achnowledged by name in the text.

S.F.J.

1 Convention

Convention is a method of marking a given circuit diagram to provide an unambiguous link between the circuit quantities and the mathematical or graphical solution which involves them. The 'sense arrow convention' used in this book is described in this chapter.

1.1 'Sense' scalars

In their work the engineer and technician are concerned with measuring distances, volumes, weights, resistances, currents, forces etc., and these quantities are classified as either SCALARS or VECTORS. The SCALAR quantity is that which is completely specified by its magnitude. For instance: distance of 5 miles or a resistance of 5 ohms. The VECTOR quantity is one which possesses:

  (i) magnitude,
 (ii) direction,
(iii) sense.

For example: the 'force' shown in Fig. 1.1 is a vector quantity, because it has a *magnitude* of 100 N, specified numerically and by the length of the line *ab*, the *direction* indicated by the position of the line *ab* in the plane of the paper, and the *sense* shown by an arrow which defines its action from point *a* to point *b* along the line. Other examples of vector quantities in electrical engineering are: Magnetic Flux Density, Electric Flux Density, and Electric Current Density.

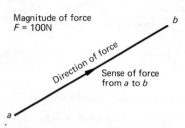

Fig. 1.1   Force as a vector

When an e.m.f. or potential difference or an electric current are considered, it is clear that they cannot be specified by their magnitude alone, because they also have a definite sense or 'direction' within the circuit.

For example, in the case of an e.m.f. it is necessary to state the polarity of the battery producing it, or for the current to indicate the 'direction' of its flow in a conductor.

Yet clearly neither the e.m.f. of the battery nor the current are vectors, for the positioning of the battery or placing of the conductor on the bench does not affect their direction as it would when forces were to be measured. Therefore, e.m.f., p.d., and the current may properly be described by means of:

 (i) magnitude,
 (ii) sense—incorrectly but commonly referred to as 'direction'.

Such quantities, therefore, should correctly be called SENSE SCALARS to differentiate them from ordinary scalar quantities. In these cases it is usual to denote their sense, i.e. 'direction', by means of an algebraic sign placed in front of their numerical values.

The *linking* of 'direction' as given by an algebraic sign to 'direction' of the quantity in the circuit diagram is done in this book by means of a 'sense arrow' convention which is described below. The 'sense' will be referred to as 'direction' but it must not be confused with the meaning of the word as applied to vector quantities.

1.2   'Sense' arrows

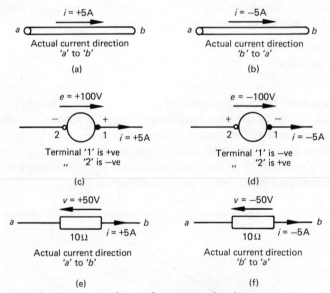

Fig. 1.2   Sense arrows for e.m.f., current and p. d.

In Fig. 1.2 (a) the conventional current of 5 amperes is assumed to flow from *a* to *b* in the conductor (the electrons flow in the opposite direction). If the sense arrow is placed pointing from *a* to *b* then the current is specified as being positive, that is + 5 A. When the current reverses and flows from *b* to *a* as in Fig. 1.2 (b), then its value changes to minus: that is, −5 A, although the sense arrow still points from *a* to *b*.

In Fig. 1.2 (c) the source of an e.m.f. is shown. When the e.m.f. of say 100 V acts from terminal 2 towards terminal 1, driving a current i in that direction, then the arrow is placed pointing towards 1 and the e.m.f. is equal to +100 V. Again, when the e.m.f. reverses and acts now from 1 to 2, the arrow is not reversed but the algebraic sign changes to minus, giving an e.m.f. of −100 V (Fig. 1.2 (d)).

Similarly in Fig. 1.2 (e) the current of 5 A flowing through a resistance of 10 Ω from a to b produces a potential difference of 5·10 = 50 V, with point a being at a higher potential than b. The arrow pointing towards a thus indicates a positive sense of voltage rise, that is +5·10 = +50 V. In Fig. 1.2 (f) the current reverses, which results in change of algebraic sign for the p.d. from plus to minus, but the arrow's direction remains unchanged.

In general therefore:

(a) The sense arrow pointing in the same direction as the actual current flow indicates a positive sense of the current and its numerical value is plus.

(b) When the sense arrow points towards terminal of higher potential for e.m.f. or a potential difference, then their sense is positive and the numerical values are also plus.

(c) The reversal of the current, e.m.f. or potential difference *does not* reverse the direction of the sense arrow but merely changes the algebraic sign. This property is very important for it enables the sense arrows to be used not only for d.c. circuits but also for labelling a.c. circuits.

(d) The sense arrows for current, e.m.f., or p.d., agree in direction for a source of e.m.f. and oppose each other in the case of a resistance. Therefore the convention gives an indication as to which circuit element supplies the electrical energy and which consumes it, that is, acts as the load.

1.3   Rules for placing sense arrows

(1) DIRECT CURRENT QUANTITIES
These are represented by CAPITAL letters.

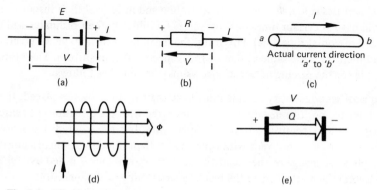

Fig. 1.3   Direct current sense arrows

1. *E.m.f.* − *E*. The sense arrow indicating the e.m.f. of a source must be placed in such a way that its tip points towards the positive terminal of the source (Fig. 1.3 (a)).

2. *Potential difference – V.* The sense arrow indicating the potential difference of a source, or across a load element of the circuit, must be placed in such a way that its tip points towards the positive terminal of the source or element (Fig. 1.3 (b)). This rule is exactly the same as for an e.m.f. It is to be noted that the sense arrow for V always points in the opposite direction to that of the current sense arrow across a load element, but in the same direction as a current sense arrow for a source of an e.m.f.

3. *Current – I.* The sense arrow indicating electric current must be placed so that its direction agrees with the direction of motion of positive electricity, i.e., the direction of conventional current (Fig. 1.3 (c)).

4. *Magnetic flux – Φ.* The sense arrow indicating magnetic flux must agree with the direction of the direct current as it progresses through the coil.

The coil is thus assumed to be wound in a clockwise direction as viewed from the terminal through which the current enters it (Fig. 1.3 (d)).

5. *Electric flux – Q.* The sense arrow indicating an electric flux Q, always points from the plate with positive potential towards the plate with negative potential. Thus its arrow opposes the potential difference arrow V.

An example of the use of sense arrow notation in a d.c. circuit is given in Fig. 1.4.

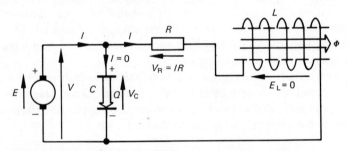

Fig. 1.4   Sense arrows in d.c. circuit

(II) ALTERNATING CURRENT QUANTITIES

These are represented by lower case (small) letters, and their r.m.s. values by capital letters. Quantities $e, v, i, Φ$ and q, vary with time and in general their instantaneous values will have at certain times positive values (graphically their ordinates will lie above the horizontal time axis). Physically the meaning of instantaneous positive values of these functions is the same as that of the respective d.c. quantities. Therefore, the general rule for the placing of 'sense' arrows may be stated as follows:

a.c. quantities which vary in time must have their sense arrows so placed, that their directions for positive instantaneous values agree with the directions of sense arrows for the respective constant d.c. quantities. It is to be noted that sense arrows so placed do not vary with time, that is their directions remains constant and a physical change of the quantity which they represent is noted by a change of the algebraic sign attached to the instantaneous value of the quantity.

From the above, particular rules follow for:

1. *E.m.f. – e.* The sense arrow indicating the e.m.f. of a source must be so placed that its tip points always towards the positive terminal for positive instantaneous values of

the e.m.f. In Fig. 1.5 the terminal 1 is positive for instantaneous values of an e.m.f. during time interval $O - t_1$ and $t_2 - t_3$, whereas terminal 2 is positive during time interval $t_1 - t_2$ and $t_3 - t_4$. The sense arrow, however, points towards terminal 1 during the whole time interval $O - t_4$.

2. *Potential difference – v.* The sense arrow indicating potential difference must be placed in exactly the same way as for an e.m.f., that is its tip points always towards the positive terminal for positive instantaneous values of the potential difference (Fig. 1.5).

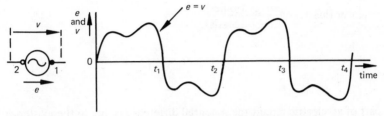

Fig. 1.5 Alternating e.m.f. and potential difference sense arrows

3. *Current – i.* The sense arrow indicating a current must be so placed that its tip points in the direction of flow of conventional current for all positive instantaneous values of that current. In Fig. 1.6 the current flows from *a* to *b* during time intervals $O - t_1$ and $t_2 - t_3$, whereas during time intervals $t_1 - t_2$ and $t_3 - t_4$ it flows from *b* to *a*. The sense arrow, however, points always in the direction $a \rightarrow b$.

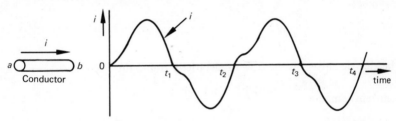

Fig. 1.6 Alternating current sense arrow

4. *Magnetic flux – Φ.* The sense arrow must point in the direction which agrees with the flux direction for all its positive instantaneous values.

5. *Electric flux – q.* The sense arrow must point in the direction which agrees with the flux direction for all its positive instantaneous values.

An example of the use of sense arrow notation in a.c. circuit is given in Fig. 1.7.

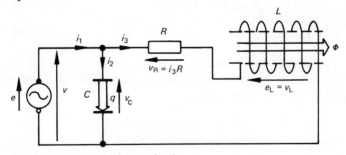

Fig. 1.7 Sense arrows in a.c. circuit

It is to be noted that the directions of sense arrows are in every respect the same as in the case of the d.c. circuit shown in Fig. 1.4.

1.4   Power in electric circuits

The power, in general, is defined as the amount of energy either produced by, or received by, a device in unit time.

Thus: power (watts) = $\dfrac{\text{energy (joules)}}{\text{time (seconds)}}$

or in symbols:

$$P = \frac{W}{t} \tag{1.1}$$

In any part of an electric circuit the potential difference is due to the existence of an electro-motive force somewhere in that circuit. The p.d. in turn is regarded as the agent that causes the current to flow in its part of the circuit. As such, it is measured by the amount of energy it imparts to the quantity of electricity in motion.

$\therefore$ potential difference (volts) = $\dfrac{\text{energy (joules)}}{\text{electricity (coulombs)}}$

or in symbols:

$$V = \frac{W}{Q} \tag{1.2}$$

The electric current is the flow of electricity round the circuit.

$\therefore$ current (amperes) = $\dfrac{\text{electricity (coulombs)}}{\text{time (seconds)}}$

or in symbols:

$$I = \frac{Q}{t} \tag{1.3}$$

Substituting from equations (1.2) and (1.3) for energy and time respectively,

$$P = \frac{W}{t} = \frac{VQ}{Q/I} = VI$$

$$\therefore P = VI \tag{1.4}$$

Thus the power in an electric circuit is equal to the product of p.d. and current.

In alternating current circuits the p.d. and current vary in time and therefore their product gives an 'instantaneous' power, i.e.,

$$p = vi \tag{1.5}$$

In Fig. 1.8(a) a simple d.c. circuit and in Fig. 1.8(b) a single phase a.c. circuit are shown, both marked according to the convention.

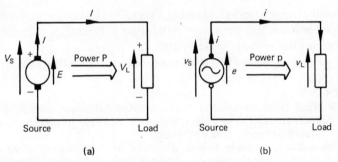

Fig. 1.8 Power in d.c. and a.c. circuits

In each case the source is identified by p.d. and current arrows in the same direction, and the load by a p.d. arrow opposing the current arrow.
The power supplied from:

(i) a d.c. source $P = V_s I$ (W)
(ii) an a.c. source $p = v_s i$ (W)

and the power received by:

(iii) a d.c. load $P = V_L I$ (W)
(iv) an a.c. load $p = v_L i$ (W)

If all the products (i) to (iv) are positive then the elements are in fact as indicated by the arrows, that is the sources supply, and the loads receive the electrical energy, and the power flow is shown by a double line arrow.
If any one of the four products becomes negative giving 'negative' power, then the function of the element is opposite to that indicated by the two arrows on the circuit diagram, i.e., the source is in fact a load and the load is a source of electrical energy. Thus the convention readily identifies the active (supply) and passive (load) elements in both a.c. and d.c. circuits.

1.5   Per unit efficiency

The efficiency of any device is determined by comparing the power input with the power output, therefore:

$$\text{efficiency } \eta = \frac{\text{Output power}}{\text{Input power}} \text{ , per unit value} \tag{1.6}$$

The difference between the two quantities is the loss incurred in the device itself. The equation (1.6) can thus be transformed into various forms as follows:

$$\eta = \frac{\text{input} - \text{loss}}{\text{input}}$$

$$= 1 - \frac{\text{loss}}{\text{input}}$$

or

$$\text{efficiency } \eta = 1 - \frac{\text{loss}}{\text{output} + \text{losses}} \text{ , per unit value} \tag{1.7}$$

The value obtained from (1.6) and (1.7) is referred to as 'per unit' because it is based on input as unity, and will always be less than 1. Percentage efficiency is obtained by multiplying the 'per unit' value by 100.

In general, equation (1.7) gives more accurate results than equation (1.6).

SUMMARY

1. Electrical quantities such as e.m.f. potential difference, current, and magnetic and electric flux, are NOT true scalars, because in addition to magnitude they also have 'sense', loosely referred to as a 'direction'.

2. This makes it necessary to link 'direction' of each quantity in the circuit to its numerical value. For this purpose, a 'sense arrow convention' is used in this book.

3. Power in the electric circuit is given by the product of p.d. and current,

i.e., for d.c. circuit $P = VI$ (W)   (1.4)

for a.c. circuit $p = vi$ (W)   (1.5)

4. The convention as described, applies to a.c. and d.c. circuits including electrical machines as circuit elements and also indicates the direction of power flow.

5. Efficiency $= \dfrac{\text{output}}{\text{input}} = 1 - \dfrac{\text{loss}}{\text{output} + \text{loss}}$, per unit.   (1.7)

Review of circuit theory

Electric currents are either DIRECT or ALTERNATING. The DIRECT current is the flow of electrons in ONE UNCHANGING direction (sense) round the electric circuit.

The ALTERNATING current is the flow of electrons first in one direction (sense) and then in the reverse direction (sense) round the electric circuit. The flow in this case is in a series of fluctuations. Furthermore, the alternating current may be either single phase or polyphase. In the latter case the most common system in use today is the three-phase one.

In this Chapter a review is given of:

   (i) direct current circuit theory;
   (ii) single phase alternating current circuit theory;
   (iii) three-phase alternating current circuit theory.

The fluctuations of an alternating current may take many forms, which when plotted as a graph of current against time produce waves of various shapes.

Of these, sinewave has been adopted as the basic shape for generation and utilisation of electrical energy, and therefore, the theory of (ii) and (iii) is confined to sinusoidal waveforms only.

DIRECT CURRENT CIRCUIT THEORY

2.1   Direct current

The feature which distinguishes direct from alternating current, is its unchanging direction (sense) of flow round the circuit. Figure 2.1(a) shows a simple electric circuit in which the 'conventional' current flows from the positive terminal of a source, in a clockwise direction, through the load and back to the negative terminal.

The direction of a conventional current is opposite to the actual direction in which the electrons flow.

When the instantaneous values of current measured at regular intervals by an ammeter A, are plotted against time, the resulting graph shown in Fig. 2.1(b) is obtained. Since the direction of the current flow does not change, all the values are positive and the graph lies wholly above the time axis. The height of the graph shown is constant, which is usually the case during 'steady state' operation, that is, between 'switching ON and OFF' times. However, this need not be so, for as long as the direction round the circuit does not change, even though its magnitude may vary, it still remains a *direct current*.

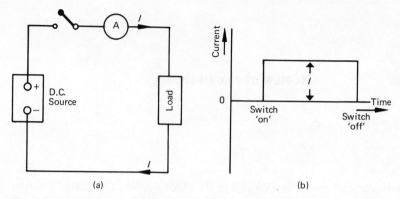

(a)   (b)

Fig. 2.1   D.C. circuit and current graph

2.2   Ohm's Law

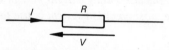

Fig. 2.2   Direct current flowing through a resistance

Figure 2.2 shows a resistance of R ohms, carrying a current of I amperes. The potential difference across its terminals is V volts. Ohm's Law states that these three quantities are connected by an equation:

$$I = \frac{V}{R}$$
(2.1)

provided the units are as specified above.

2.3   Kirchhoff's Laws

The First (current) Law states:
  'The algebraic sum of currents in conductors meeting at a junction is equal to zero', i.e.:

$$\Sigma I = 0$$
(2.2)

The Second (voltage) Law states:
  'In any closed loop of a network the algebraic sum of potential differences and e.m.f's of that loop is equal to zero', i.e.:

$$\Sigma E + \Sigma IR = 0$$
(2.3)

The current law is illustrated in Fig. 2.3(a) where four conductors carrying current I_1, I_2, I_3 and I_4 amperes, meet at the junction J. The sense arrows indicate the current directions, and assigning positive signs to currents flowing towards, and a negative sign to a current flowing away from the junction, the equation:

$$I_1 + I_2 - I_3 + I_4 = 0$$

is obtained.

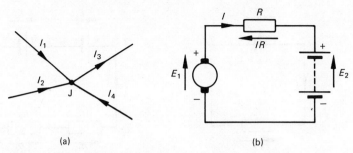

Fig. 2.3 **Circuits illustrating Kirchhoff's Laws. (a) current law
(b) voltage law**

The convention of attaching positive signs to currents flowing towards any junction of the network, and negative signs to currents flowing away from that junction, is adhered to in this textbook.

As an illustration of the voltage law, consider the circuit in Fig. 2.3(b). The polarities of a generator and a battery are indicated by 'sense' arrows, placed next to E_1 and E_2. If E_1 is greater than E_2, then the current I flows round the circuit in a clockwise direction. The sense arrow indicating potential difference across the resistor must point towards the positive terminal of the generator, according to the convention (i.e. the potential sense arrows are always directed towards the points at higher i.e. positive potential). It is to be noted that potential difference sense arrows on the load elements (in this case the resistor) always oppose the current sense arrows. By going round the circuit in a clockwise direction and assigning positive algebraic signs to the arrows, the directions of which are consistent with the motion, and negative signs to arrows which oppose the motion, the equation:

$$E_1 - IR - E_2 = 0 \text{ or } E_1 = IR + E_2$$

is obtained.

By going round the circuit in an anti-clockwise direction and assigning algebraic signs as described above, the same equation:

$$E_2 + IR - E_1 = 0 \text{ or } E_1 = IR + E_2$$

is obtained.

Obviously in forming the equation, the choice of direction round the loop is immaterial.

2.4   Series circuit

Figure 2.4(a) shows the simplest circuit consisting of a battery of negligible internal resistance supplying a resistor. Applying Kirchhoff's voltage law,

$$E - IR = 0 \text{ or } E = IR$$

is obtained, which is, of course, Ohm's Law.

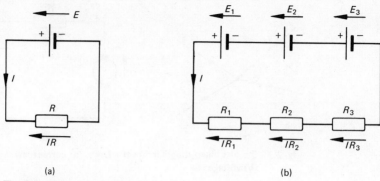

(a)   (b)

Fig. 2.4   Series d.c. circuits

Considering now the circuit in Fig. 2.4(b), the sense arrows are first placed according to the convention. The application of Kirchhoff's Law then yields the equation:

$$E_1 + E_2 + E_3 - IR_1 - IR_2 - IR_3 = 0$$

from which,

$$I = \frac{E_1 + E_2 + E_3}{R_1 + R_2 + R_3} = \frac{E_t}{R_t}$$

where $E_t = E_1 + E_2 + E_3$

and $R_t = R_1 + R_2 + R_3$

The above equations give the required current and show that:

(i) the total e.m.f. E_t of batteries connected in series in the same direction is equal to the sum of the e.m.f.'s of each cell, i.e.,

$$E_t = E_1 + E_2 + E_3 \qquad\qquad (2.4)$$

and

(ii) that resistances in series may be replaced by one resistor R_t the magnitude of which is equal to the sum of all the resistances, i.e.,

$$R_t = R_1 + R_2 + R_3 \qquad\qquad (2.5)$$

QUESTION 1. Two batteries of negligible internal resistance are connected in series. Their e.m.f.'s are 40 V and 50 V respectively. They supply three resistors of 5 Ω, 10 Ω and 15 Ω connected in series. Find the current in the circuit and the reading of a voltmeter, which has one lead connected to the junction of the two batteries, and the other lead to the junction of the 5- and 10-Ω resistors. Assume the resistance of the voltmeter to be infinite.

ANSWER

The first diagram represents the arrangement of the circuit elements with the sense arrows placed as described previously. The voltage equation obtained from it is:

$$50 + 40 - 5I - 10I - 15I = 0$$

from which

$$I = \frac{90}{30} = 3 \text{ A}$$

After re-drawing the circuit with actual figures for potential differences across the resistors, the voltmeter reading is obtained by considering the p.d. sense arrows:

$$V \text{ (Voltmeter)} = 40 - 15 = 25 \text{ V}$$

or,
$$V \text{ (Voltmeter)} = 45 + 30 - 50 = 25 \text{ V}$$

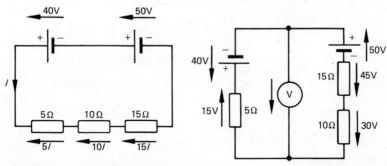

Fig. 2.5 Circuit diagrams for Question 1

Since the answer is positive, the direction of the voltmeter sense arrow is correct and the junction of 5- and 10-ohm resistors is at a higher potential with respect to the junction of the batteries.

2.5   Parallel circuit

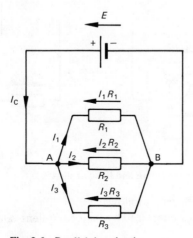

Fig. 2.6 Parallel d.c. circuit

The parallel circuit shown in Fig. 2.6 can be treated as follows:

(i) Kirchhoff's current law is applied to point A, which gives

$$I_t = I_1 + I_2 + I_3.$$

Consideration of point B gives an identical equation.

(ii) Kirchhoff's voltage law yields:

$$E = I_1 R_1 = I_2 R_2 = I_3 R_3$$

therefore,

$$I_1 = \frac{E}{R_1}, I_2 = \frac{E}{R_2} \text{ and } I_3 = \frac{E}{R_3}.$$

(iii) Substituting for currents in the first equation

$$I_1 = \frac{E}{R_1} + \frac{E}{R_2} + \frac{E}{R_3} = E\left(\frac{1}{R_1} + \frac{1}{R_2} + \frac{1}{R_3}\right) = E\left(\frac{1}{R_e}\right) = \frac{E}{R_e}$$

where R_e is the equivalent resistance of the parallel network.

The above gives the solution to the network from first principles and proves the relation:

$$\frac{1}{R_e} = \frac{1}{R_1} + \frac{1}{R_2} + \frac{1}{R_3} \qquad (2.6)$$

for resistances in parallel.

2.6   Series – parallel circuit

Figure 2.7(a) shows a simple series-parallel circuit

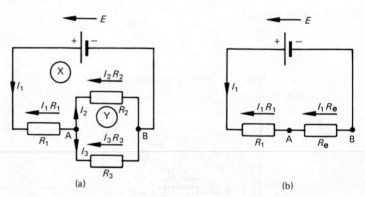

Fig. 2.7   Series-parallel circuit and its equivalent

The solution of the network is obtained as follows:

(i) The parallel branch resistances are replaced by their equivalent value R_e which is obtained from equation (2.6) i.e.,

$$\frac{1}{R_e} = \frac{1}{R_2} + \frac{1}{R_3}$$

therefore $$R_e = \frac{R_2 \times R_3}{R_2 + R_3}$$

This reduces the circuit to that of Fig. 2.7(b)

(ii) Using the simplified circuit, Kirchhoff's voltage law gives:

$$E - I_1 R_1 - I_1 R_e = 0$$

Hence $\qquad I_1 = \dfrac{E}{R_1 + R_e}$, where $(R_1 + R_e)$ is the total resistance of the circuit.

Since points A and B in Fig. 2.7(a) and 2.7(b) are identically placed with respect to the portion of the circuit containing the battery and resistance R_1, hence:

$$I_1 R_e = I_2 R_2 = I_3 R_3,$$

where $I_2 R_2 = I_3 R_3$ is also a consequence of Kirchhoff's Second Law applied to the loop Y in Fig. 2.7(a).

Thus $\qquad I_2 = \dfrac{I_1 R_e}{R_2}$ and $I_3 = \dfrac{I_1 R_e}{R_3}$

QUESTION 2. In Fig. 2.7(a) $E = 54$ V, $R_1 = 6\ \Omega$, $R_2 = 8\ \Omega$, $R_3 = 12\ \Omega$. Find all currents and voltages.

ANSWER

$$R_e = \frac{R_2 \times R_3}{R_2 + R_3} = \frac{8 \times 12}{8 + 12} = 4 \cdot 8\ \Omega$$

and $\qquad I_1 = \dfrac{54}{6 + 4 \cdot 8} = 5$ A

Potential difference between A and B is equal to:

$$R_e I_1 = 5 \times 4 \cdot 8 = 24\ V$$

whence $\qquad I_2 = \dfrac{24}{8} = 3$ A.

and $\qquad I_3 = \dfrac{24}{12} = 2$ A

The complete solution is given in the form of a diagram below. The sense arrows are directed at the points of higher potential i.e., plus (+).

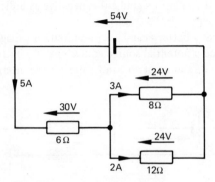

Fig. 2.8

In all the cases considered above, the source of e.m.f. (battery) has been assumed to have no internal resistance, i.e., the source was capable of supplying any load current without any change in its terminal p.d., which was always equal to an e.m.f. E. Sometimes an assumption of this kind may be made, but in the majority of cases a d.c. source cannot be represented simply by its constant e.m.f.

2.7   Equivalent circuit of a d.c. source

Suppose the following experiment is performed. A battery is connected across a variable resistor in series with the ammeter as shown in Fig. 2.9(a). The voltmeter is connected across the terminals of the battery and the readings of the voltmeter and ammeter are taken for different values of the variable resistor R.

It is assumed that the voltmeter resistance is infinite, i.e., the voltmeter takes no current, and that the ammeter resistance is negligible, so that the potential difference across it is zero. The graph of the battery's terminal voltage V is then plotted in Fig. 2.9(b) against the current delivered to the resistor. It is found that this gives a straight line, sloping from left to right.

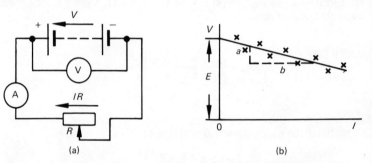

(a)   (b)

Fig. 2.9   Circuit and graph to determine an equivalent potential source of a battery

The equation of the resulting line can be written as

$$V = E - r \cdot I \tag{2.7}$$

where E is an intercept on the voltage axis and $r = \dfrac{a}{b}$ is the gradient of the line.

Since the line slopes from left to right its gradient is negative as indicated by a minus sign in front of r.

The equation (2.7) shows that a battery can be represented on a diagram by the circuit of Fig. 2.10. This is called a potential source equivalent of a battery, where E is an internal potential source, i.e., source of an e.m.f., and r is the internal resistance of the source.

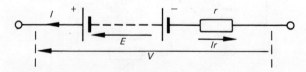

Fig. 2.10   Potential source equivalent of the battery

Sources for which the potential/current relationship can be written as a linear equation, are called linear sources. Similarly, the potential/current relationship for a d.c. machine may be represented approximately by a linear equation:

$$V = E_r \pm I_a R_a \tag{2.8}$$

where E_r is an e.m.f. of rotation, I_a armature current, R_a resistance of the commutator winding, and V is the p.d. across the winding. Hence, on the basis of equation (2.8) a d.c. commutator winding can be represented as in Fig. 2.11(a) and (b).

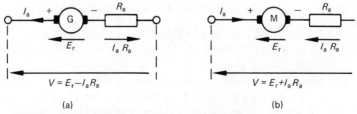

(a)   (b)

Fig. 2.11 Potential source equivalent of a d.c. machine (a) Generator
(b) Motor

Summarising, it is seen that in circuit analysis batteries, d.c. machines, and other linear potential sources can be regarded as being made up of two elements: (i) a source of e.m.f. E, which is independent of the load current I, and (ii) *internal* resistance r.

QUESTION 3. It is required to find the e.m.f. and the internal resistance of a battery. The following two tests were performed:

1. Resistance A was connected across the battery and readings of current and terminal p.d. were found to be 3 A and 19·2 V.
2. Resistance A was then replaced by another resistance B, the readings obtained in this case being 2 A and 20 V respectively.

ANSWER. The battery is assumed to be a linear source, hence two equations are formed:

$$19{\cdot}2 = E - 3r \tag{i}$$
and
$$20 = E - 2r \tag{ii}$$

The solution of these equations gives $E = 21{\cdot}6$ V and $r = 0{\cdot}8\ \Omega$. Thus the potential source equivalent of the battery is expressed by an equation:

$$\underline{V = 21{\cdot}6 - 0{\cdot}8I \text{ (V)}}$$

2.8   More complex circuits

Previous sections deal with circuits, the solutions of which are obtained by forming an equation with one unknown, usually a current. Sometimes it is necessary to replace parallel branches by an equivalent resistor, before such an equation can be obtained, but in general only one equation is necessary.

In this section, networks considered are such that it is not possible to resolve them into a single loop circuit through application of the relationship:

$$\frac{1}{R_e} = \frac{1}{R_1} + \frac{1}{R_2}$$

or

$$R_t = R_1 + R_2 + R_3 + \ldots$$

Hence two or more simultaneous equations will be necessary before the solution can be found. These equations are formed by a consistent application of both Kirchhoff's Laws.

The procedure to be followed is summarised below:

(i) the circuit diagram is drawn;

(ii) all values of resistances and known e.m.f.'s are inserted, including the latter's sense arrows;

(iii) using Kirchhoff's First Law all currents with their assumed directions are drawn in. The number of currents must be kept to a minimum.

(iv) All voltage arrows are inserted next to every circuit element, remembering that on the resistance the current and voltage arrows oppose each other, whilst for sources of e.m.f. their directions agree;

(v) equations are formed by applying Kirchhoff's Second Law and attaching positive signs to voltage arrows which agree with the direction of progression around each loop, and negative signs to those which oppose it;

(vi) the equations are solved;

(vii) the current and voltage arrows are corrected, whenever the answers to assumed currents and voltages are *negative.*

Question 4 illustrates the above.

QUESTION 4. A circuit is shown in Fig. 2.12(a). G is a permanent magnet generator producing an e.m.f. of 12 V, its armature circuit resistance being 0·5 Ω. B is a battery of e.m.f. equal to 6 V and internal resistance of 2 Ω. R equals 1 Ω and represents the load. It is required to find currents and the terminal p.d.s of the e.m.f. sources, when:

(i) the generator and battery are connected as in Fig. 2.12(a)

and (ii) when the battery connections are inadvertently reversed as in Fig. 2.12(c)

ANSWER

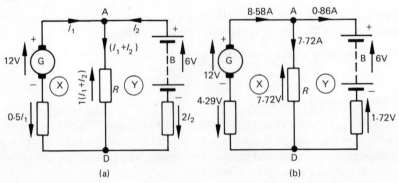

(a)   (b)

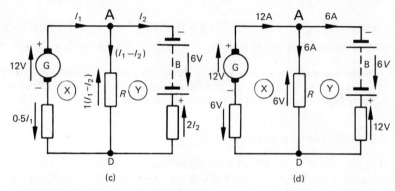

Fig. 2.12   Circuit diagrams for Question 4

Current sense arrows I_1 and I_2 are placed in an arbitrary manner in Fig. 2.12(a) and assuming current direction through 1 Ω resistor to be from A to D, the application of Kirchhoff's current law gives:

$$I_1 + I_2 - (\text{current in branch } AD) = 0$$

therefore current in $AD = (I_1 + I_2)$

The voltage sense arrows are now inserted according to the convention and applying Kirchhoff's voltage law, p.d. equations are formed.

The number of unknowns determines the number of equations necessary, in this case two.

Loop X starting from the generator in a clockwise direction:

$$12 - 1(I_1 + I_2) - 0.5\,I_1 = 0$$

or, $1.5\,I_1 + I_2 = 12$ (i)

Loop Y starting from the battery in an anticlockwise direction:

$$6 - 1(I_1 + I_2) - 2\,I_2 = 0$$

or $I_1 + 3\,I_2 = 6$ (ii)

Solution of the equations (i) and (ii) gives

$$I_1 = 8.58 \text{ A}$$

and $I_2 = -\,0.86 \text{ A}$

Hence current I_2 actually flows towards the positive terminal of the battery i.e., the battery is being charged. Figure 2.12(b) shows the circuit with actual values placed next to sense arrows for currents and potential drops. It is clear from the potential arrows that the p.d. between points A and D is given by

$$(12 - 4.29) \simeq 7.72 = (6 + 1.72) \text{ V.}$$

The slight discrepancy in the last figure is due to working to two decimal places for current values.

When the polarity of the battery is reversed the diagram is as shown in Fig. 2.12(c). The new equations are:

Loop X. $12 - 0.5\,I_1 - 1(I_1 - I_2) = 0$

$1.5\,I_1 - I_2 = 12$ (iii)

Loop Y. $6 - 2\,I_2 + 1(I_1 - I_2) = 0$

$-I_1 + 3\,I_2 = 6$ (iv) therefore $I_1 = \dfrac{42}{3 \cdot 5} = 12$ A and $I_2 = 1 \cdot 5\, I_1 - 12 = 6$ A. Again Fig. 2.12(d) gives a clear picture of the potential distribution. Potential difference between A and D is $12 - 6 = 6 = 12 - 6$ V.

Note that the potential difference across the internal resistance of the battery is twice as large as the battery's e.m.f. Hence the negative terminal of the battery will actually be (+) with respect to its normally positive terminal.

2.9   Power in d.c. circuits

If in any part of a d.c. circuit a steady potential difference of V volts produces a constant current of I amperes through that portion of the circuit, then the power dissipated by it is P watts, i.e.:

$$P = VI \tag{1.4}$$

By substitution from Ohm's Law (2.1)

$$P = I^2 R \tag{2.9}$$

or

$$P = \frac{V^2}{R} \tag{2.10}$$

SINGLE PHASE ALTERNATING CURRENT CIRCUIT THEORY

2.10   Single phase alternating current (a.c.)

The current which changes its direction of flow periodically round the circuit is called an alternating current, and the term 'single phase' is used to describe one source of alternating current connected to a load through *two* wires. Such an arrangement is shown in Fig. 2.13(a). When the terminal 1 is positive with respect to the terminal 2 the current flows in a clockwise direction round the circuit. When the terminal 2 becomes positive with respect to the terminal 1, the current reverses its direction and flows in an anticlockwise direction.

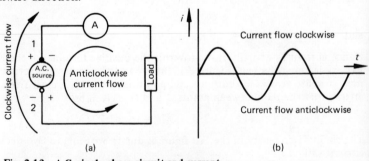

(a)   (b)

Fig. 2.13   A.C. single phase circuit and current

If a centre-zero ammeter is used to measure the current and the reversals of polarity of the source are very slow, the graph of the current against time can be plotted. Assigning positive values to the current flowing in a clockwise direction and negative to that flowing in an anticlockwise direction, the graph is obtained which is shown in Fig. 2.13(b).

The illustrated waveform of the current is called a sine wave, one half of which lies above, and the other half below the time axis.

2.11   Sinusoidal potential of a source

Consider a source of e.m.f. such that the potential difference across its terminals is given by an equation:

$$v = V_m \sin \omega t$$

For a sinewave the effective (r.m.s.) value

$$V = \frac{V_m}{\sqrt{2}}$$

$$\therefore \qquad V_m = \sqrt{2}\, V$$

and the equation may be rewritten as

$$v = \sqrt{2}\, V \sin \omega t \qquad\qquad (2.11)$$

where v is the instantaneous voltage across the terminals of the source in volts.

V is the r.m.s. value of the p.d. in volts
ω is the angular velocity in radians per second
and t is the time in seconds, so that
ωt is the angle in radians.

Let this source be represented diagrammatically in Fig. 2.14(a) whilst Fig. 2.14(b) shows the graph of the equation (2.11).

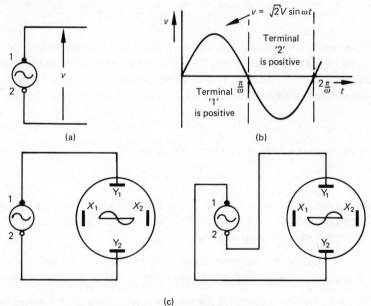

Fig. 2.14   **Appertaining to sinusoidal potential source**

When the oscilloscope is used to investigate the p.d. of the source, two ways of connecting terminals 1 and 2 to the Y plates are possible. Each connection produces a trace of a sinewave, but the phase difference between them is 180°, as shown in Fig. 2.14(c).

Thus the reversal of connections results in the trace shift of 180°. It is clear from the foregoing that some kind of link between the polarity of the source and the plot of equation representing its potential difference is necessary to avoid ambiguity. This is provided by means of a potential sense arrow. In Fig. 2.14(a) the arrow representing v points towards terminal 1, and in Fig. 2.14(b) the portion of the graph above the time axis therefore corresponds to the p.d. of the source when the terminal 1 is *positive* with respect to terminal 2, as explained in Chapter 1.

Thus a definite relation is established between the p.d. of the source and its graphical and mathematical representation. This is very important in a.c. work, especially where several sources are to be interconnected.

2.12   Kirchhoff's Laws in a.c. circuits

Consider two sinusoidal sources of e.m.f. producing potential differences v_1 and v_2 across their terminals.

The frequencies of the sources are the same. Let the sources be connected in series across a load as shown in Fig. 2.15(a).

By Kirchhoff's Law of potentials:

$$v = v_1 + v_2 \tag{2.12}$$

or

$$v = v_1 - v_2 \tag{2.13}$$

Obviously it is necessary to know the polarities of the sources in order that the correct equation may be written. Only one of the two stated above can be correct in a given case. Let the polarities of the sources be specified as in Fig. 2.15(b).

Then, when the terminals 2 and 3 are joined together, the equation

$$v = v_1 + v_2 \text{ applies, and}$$

if the terminals 2 and 4 are joined, the equation

$$v = v_1 - v_2 \text{ is valid.}$$

Both cases are illustrated in Fig. 2.15(c) and (d) where for simplicity v_1 and v_2 have been assumed to be in step or *in phase* with v_1 being greater than v_2.

The above example indicates that:

'Kirchhoff's Law for potential differences should be applied as if each potential difference had a constant direction, that is that of its sense arrow.'

In the case of currents which continually reverse their direction similar considerations apply.

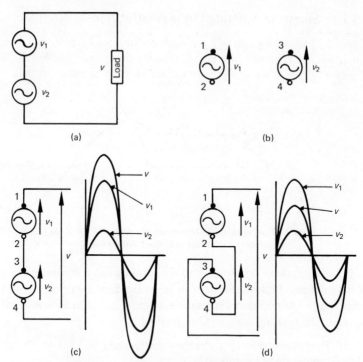

Fig. 2.15 (a) Two a.c. sources in series – polarities unspecified
(b) Polarities of the two a.c. sources specified (c) Two a.c.
sources in series equation (2.12) applies (d) Two a.c. sources
in series equation (2.13) applies

Consider a junction of three conductors carrying sinusoidal currents. The sense arrows which are placed on the conductors indicate that the actual direction of the current flow is regarded as positive if it agrees with the direction of the sense arrow (See Chapter One). Hence, in Fig. 2.16 the equation may be formed by applying Kirchhoff's Law for currents as if each current direction were constant and in agreement with its sense arrow, i.e.,

$$i = i_1 + i_2 \qquad\qquad (2.14)$$

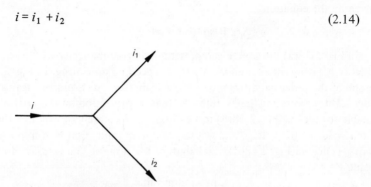

Fig. 2.16 Junction of three conductors carrying a.c. currents

2.13   Sinewave generated by a rotating line

Figure 2.17 shows a graph of a sinewave constructed by plotting vertical projections of a rotating line against time (or time angle) marked off horizontally. The angular velocity of the line is constant and equal to ω rad/s.

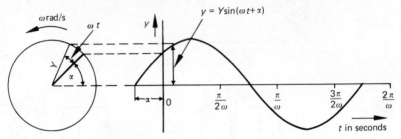

Fig. 2.17   A rotating line generating a sinewave $y = Y \sin (\omega t + \alpha)$. The heavy radial line is at the $t = 0$ position

It is clear that the 'knowledge' of the rotating line is equivalent to the knowledge of the sinewave. Hence the latter can be replaced by the former, resulting in great simplification of notation. For complete equivalence, the following information regarding the rotating line must be specified.

(a) The angular velocity (ω radians per second)
(b) The centre of rotation (O)
(c) The length (Y)
(d) The position at some known time, usually $t = 0$.

The above is done by drawing a line to scale, (length), at time $t = 0$, (position), placing an arrowhead on the moving end, (implying centre of rotation). The angular velocity, ω is placed next to the curved arrow indicating the direction of rotation of the line.

2.14   Voltage phasors

Let us consider once more a source of an e.m.f. such that its potential difference is given by the equation

$$v = \sqrt{2} V \sin (\omega t + \alpha) \tag{2.15}$$

In Fig. 2.18(a) the source is represented diagrammatically with a sense arrow placed next to v, giving the reference polarity as specified previously. In Fig. 2.18(b), the graph of the potential difference $v = \sqrt{2} V \sin (\omega t + \alpha)$ is plotted against time. In Fig. 2.18(c) a rotating line is drawn so that its projection on the vertical axis would enable the plot of Fig. 2.18(b) to be obtained. As is well known, it is the effective or r.m.s. value of the voltage and not its maximum value, that is of importance in a.c. work, hence, in Fig. 2.18(d), the length of the rotating line is scaled down to the r.m.s. value V.

For sinewaves, the ratio of $\dfrac{\text{maximum value}}{\text{r.m.s. value}} = \sqrt{2}.$

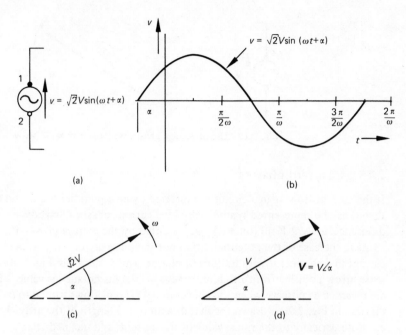

Fig. 2.18 (a) Source of a.c. e.m.f.
          (b) Graph of $v = \sqrt{2} V \sin (\omega t + \alpha)$
          (c) Rotating line which generates sinewave given by $v = \sqrt{2} V \sin$
               $(\omega t + \alpha)$ drawn at $t = 0$
          (d) Phasor symbolising $v = \sqrt{2} V \sin (\omega t + \alpha)$ drawn at $t = 0$

The line depicted in Fig. 2.18(d) thus symbolises the potential difference of the source. The name given to this line is 'phasor', and it may be denoted by V, hence,

$$V \text{ symbolises } v = \sqrt{2} \ V \sin (\omega t + \alpha) \tag{2.16}$$

The two qualities which determine a phasor quantity are its magnitude and its angle. They can be indicated by a composite symbol such as

$$V = V \angle \alpha \tag{2.17}$$

This notation makes it possible to use numerical values to specify phasors, i.e.,

$$V = 230 \angle 30° \text{ symbolises } v = \sqrt{325 \cdot 2} \sin (\omega t + 30°) \text{ V.}$$

Equation (2.17) expresses a phasor in a polar form.

It must be noted that a phasor does not vary with time, but it *symbolically* represents a time-varying function. Therefore ω is omitted from the diagram, but not its associated arrow.

The potential sense arrow, which serves as a link between the physical source and the mathematical expression of its p.d., also applies to a phasor. Figure 2.19 shows it pictorially. Note that symbol V i.e., phasor, is now placed next to the sense arrow of the source.

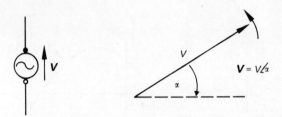

Fig. 2.19   Use of voltage phasor symbols with sense arrow notation

2.15   Current phasors

If the current flowing in a circuit is expressed by an equation $i = \sqrt{2}\,I \sin(\omega t + \beta)$, then it can be represented symbolically by a current phasor I, exactly as the potential difference, $v = \sqrt{2}\,V \sin(\omega t + \alpha)$ is represented by the voltage phasor V.

As in the case of the potential difference the sense arrow must be used to link the current in a conductor with its current phasor. This is shown in Fig. 2.20 where the sense arrow pointing from a to b indicates that the instantaneous values of the current are regarded as positive when the conventional current flows from a to b. The phasor diagram in Fig. 2.20 is drawn for that direction, the length of the arrow being made equal numerically to the r.m.s. value of the current, and inclined at an angle β to the horizontal.

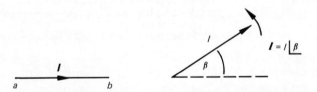

Fig. 2.20   Use of current phasor symbols with sense arrow notation

As before:

$$I = I\angle\beta \text{ symbolises } i = \sqrt{2} \sin(\omega t + \beta) \qquad (2.18)$$

or taking a numerical example:

$$I = 10\angle 15° \text{ symbolises } i = 14.14 \sin(\omega t + 15°) \text{ A}$$

2.16   Kirchhoff's Laws applied to phasors

KIRCHHOFF'S FIRST LAW

This Law, which deals with the current in conductors meeting at a point, may be stated as follows:

'The *geometric* sum of currents symbolised by phasors in conductors meeting at a point is equal to zero'

For example, consider three conductors carrying currents which are designated by three sense arrows with phasor symbols I_1, I_2 and I_3. Assigning the positive sign to a current whose sense arrow points towards junction J and the negative sign to a current whose sense arrow points away from J, the equation

$$I_1 - I_2 - I_3 = 0 \qquad\qquad (2.19)$$

is obtained.

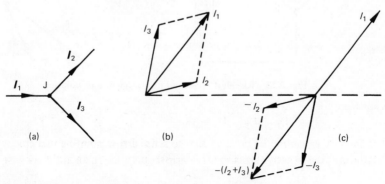

Fig. 2.21   Kirchhoff's First Law using phasor notation

Figure 2.21(b) indicates phasors of the three currents, whereas in Fig. 2.21(c) the equation is solved graphically. The negative sign in front of the *I* symbol corresponds to the rotation of a phasor through $180°$. It is clearly seen that the geometric sum is indeed zero. The above considerations are based on the fact that the sum of two sinewaves of the same frequency gives a third sine wave of a frequency equal to that of the other two, as illustrated in Fig. 2.22.

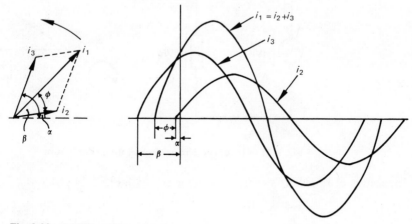

Fig. 2.22   Addition of two sinewaves

KIRCHHOFF'S SECOND LAW

'The *geometric* sum of the potential differences expressed in phasor form in any closed loop is equal to zero.'

The application of this law is shown in Fig. 2.23. The sense arrows are placed according to the convention.

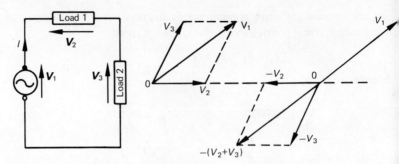

Fig. 2.23   Kirchhoff's Second Law using phasor notation

An equation

$$V_1 - V_2 - V_3 = 0 \qquad (2.20)$$

is obtained by assigning the positive sign to sense arrow pointing in a clockwise direction and the negative signs to sense arrows pointing in an anticlockwise direction.

2.17   Basic elements in a.c. circuits

1. *Resistor - R.* A non-inductive resistor of R ohms is connected across the sinusoidal source, the potential of which is given by an equation:

$$v = \sqrt{2}\ V \sin \omega t \qquad (2.11)$$

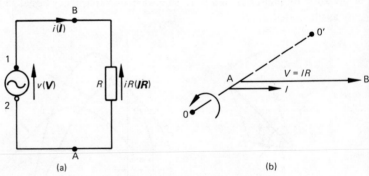

Fig. 2.24   Circuit and phasor diagram for a pure resistor

Application of Kirchhoff's voltage law to the circuit in Fig. 2.24 yields

$$v - iR = 0$$

therefore

$$i = \frac{v}{R}$$

Substituting for v from equation (2.11)

$$i = \frac{\sqrt{2}\ V \sin \omega t}{R} = \sqrt{2}\ \frac{V}{R} \sin \omega t = \sqrt{2} I \sin \omega t$$

therefore

$$i = \sqrt{2} I \sin \omega t \qquad (2.21)$$

where $I = \dfrac{V}{R}$, which re-arranged gives:

$$R = \frac{V}{I}$$ (2.22)

The above shows that a resistor in an a.c. circuit,

     (i) determines the ratio of V(r.m.s.) to I(r.m.s.)

and (ii) that the variations of voltage and current across it are *in step* with each other, or *in phase.*

It is now possible to use the above results and deal with the circuits in terms of r.m.s. values and phasors only. The circuit diagram is accordingly re-labelled with phasor symbols (shown in brackets) and the equation obtained by Kirchhoff's Second Law becomes:

$$V - IR = 0$$

or,   **$V = IR$**   (2.23)

Graphical solution of (2.23) is shown in Fig. 2.24(b), where all phasors are drawn at a time $t = 0$. It must be noted that the current phasor I is parallel to phasors V and IR, because these quantities are *in phase.*

    Since in the equation (2.23).

$$V = V\angle 0$$

and $I = I\angle 0$

multiplication of the current phasor by a resistance R *does not* change the voltage phasor's initial angle, hence an angle of $0°$ may be associated with the resistance, and may be written as $R = R\angle 0°$. R however *does not* symbolise a sinewave as V and I do.

2. *Inductance – L.* Assuming that the sinusoidal source $v = \sqrt{2}\, V \sin \omega t$ is connected across a pure inductor of L henrys, then an application of Kirchhoff's voltage law to the circuit in Fig. 2.25(a) gives:

$$v - e = 0$$

therefore $v = e,$

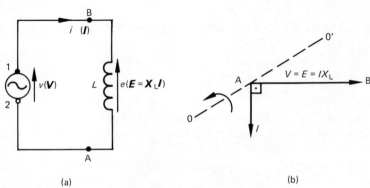

(a)   (b)

Fig. 2.25 Circuit and phasor diagram for a pure inductor

but from the fundamental considerations

$$e = L \frac{di}{dt}$$

Therefore $\qquad\qquad \sqrt{2}\, V \sin \omega t = L \frac{di}{dt}$

and $\qquad\qquad \frac{di}{dt} = \frac{\sqrt{2}\, V \sin \omega t}{L}$

Integrating, $\qquad i = -\sqrt{2}\,\frac{V}{\omega L} \cos \omega t = -\sqrt{2}\, I \cos \omega t$, or

$$i = -\sqrt{2}\, I \cos \omega t \qquad\qquad\qquad\qquad (2.24)$$

The constant of integration is omitted since initial conditions may be chosen so as to make it equal to zero.

Also, $\qquad\qquad I = \frac{V}{\omega L}$

which re-arranged gives:

$$\omega L = \frac{V}{I} = 2\pi f L = X_L \qquad\qquad\qquad\qquad (2.25)$$

X_L is called an inductive reactance.

A comparison of the current equation (2.24) with the initial voltage equation indicates that the current wave lags the voltage by $90°$. Equation (2.25) shows that the ratio of V(r.m.s.) to I (r.m.s.) is determined by an inductive reactance $X_L = 2\pi f L$. Using these two facts the circuit is dealt with in terms of r.m.s. values and phasors only. The circuit diagram is re-labelled with phasor symbols (shown in brackets) and the equation is directly obtained from it:

$$V - E = 0$$

therefore:

$$V = E = IX_L \qquad\qquad\qquad\qquad (2.26)$$

Graphical solution of (2.26) is shown in Fig. 2.25(b) with the current phasor I lagging voltage phasor V by $90°$.

Since $\qquad\qquad V = V\angle 0°$

and $\qquad\qquad I = I\angle -90°$

multiplication of current phasor by an *inductive* reactance X_L turns it through $90°$ in an *anticlockwise* direction. Hence the angle of $+90°$ may be associated with it. The composite symbol is thus

$$X_L = X_L \angle 90°,$$

so that $\qquad V\angle 0° = I\angle -90 \times X_L \angle 90 = IX_L \angle -90 + 90 = IX_L \angle 0°$

3. *Capacitance - C.* A sinusoidal source, which generates $v = \sqrt{2}\,V \sin \omega t$ is connected across a pure capacitor of C farads as in Fig. 2.26(a)

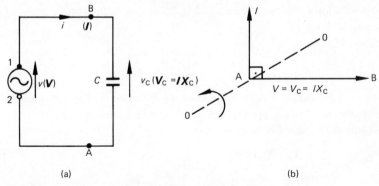

(a)   (b)

Fig. 2.26   Circuit and phasor diagram for a capacitor

Application of Kirchhoff's voltage law gives:

$$v - v_c = 0$$

therefore $v = v_c$

But from the fundamental considerations

$$i = C\frac{dv_c}{dt}$$

∴ $$i = C\frac{dv}{dt} = C\frac{d}{dt}\left(\sqrt{2}\,V \sin \omega t\right) = \sqrt{2}\,\omega C V \cos \omega t$$

$$= \sqrt{2}\,I \cos \omega t, \text{ or}$$

$$i = \sqrt{2}\,I \cos \omega t \tag{2.27}$$

where $I = \omega C V$, which re-arranged gives:

$$\frac{1}{\omega C} = \frac{V}{I} = \frac{1}{2\pi f C} = X_c \tag{2.28}$$

X_c is called a capacitive reactance.

A comparison of the current equation (2.27) with the initial voltage equation indicates that the current wave leads the voltage wave by 90°. Equation (2.28) shows that the ratio of V(r.m.s.) to I(r.m.s.) is determined by a capacitive reactance

$$X_c = \frac{1}{2\pi f C}$$

Using these two facts the circuit is dealt with in terms of r.m.s. values and phasors only. The circuit diagram is re-labelled with phasor symbols (shown in brackets) and the equation is directly obtained from it.

$$V - V_c = 0$$

therefore:

$$V = V_c = IX_c \tag{2.29}$$

Graphical solution of (2.29) is shown in Fig. 2.26(b) with the current phasor I *leading* voltage phasor V by 90°.

Since $V = V\angle 0$

and $I = I\angle 90°$

multiplication of current phasor by a *capacitive* reactance X_c turns it through 90° in a *clockwise* direction. Hence the angle of $-90°$ may be associated with it.

The composite symbol is thus:

$$X_c = X_c \angle -90°$$

so that

$$V\angle 0° = I\angle 90° \times X_c \angle -90° = IX_c \angle 90° -90° = IX_c \angle 0°.$$

Again X_c does not symbolise the sinewave.

2.18   Series circuit containing R, L and C

An a.c. source, the terminal potential of which is $v = \sqrt{2}\, V \sin (\omega t + \phi)$, supplies the current $i = \sqrt{2}\, I \sin \omega t$ to the circuit containing resistance R, inductance L and capacitance C. The circuit is drawn in Fig. 2.27(a) and labelled with phasor symbols:

$$
\begin{aligned}
I &= I\angle 0° && \text{and symbolises } i && = \sqrt{2}\, I \sin \omega t\\
V &= V\angle \phi° && \text{,,} && v && = \sqrt{2}\, V \sin (\omega t + \phi)\\
V_R &= V_R \angle 0° && \text{,,} && v_R && = \sqrt{2}\, V_R \sin (\omega t)\\
V_L &= V_L \angle 90° && \text{,,} && v_L && = \sqrt{2}\, V_L \sin \left(\omega t + \frac{\pi}{2}\right)\\
V_C &= V_C \angle -90° && \text{,,} && v_C && = \sqrt{2}\, V_C \sin \left(\omega t + \frac{\pi}{2}\right)
\end{aligned}
$$

Junctions between various elements of the circuit are labelled A, B, C and D, starting with A at the terminal of a.c. source from which sense arrow V points *away*.

Using the current I as a reference phasor, drawing it at $t = 0$ and starting with point A, the voltage phasor diagram is drawn in Fig. 2.27(b). It may be called a 'string' diagram because the phasors are added to each other 'end to end' and all the points on it correspond exactly with the points in the circuit.

The diagram graphically represents the equation:

$$V = V_R + V_L + V_C \tag{2.30}$$

Phasor V_R is drawn from A to B in phase with the current phasor I, V_L is drawn from B to C leading I by 90°, and V_C is drawn from C to D downward because it lags I by 90°.

From point D back to A the phasor V as the supply voltage, completes the 'string'. Thus all points on the circuit diagram find their counterparts on the voltage 'string' phasor diagram, and their directions (arrowheads) agree with the directions of sense arrows. The current phasor I is used only as a reference. Figure 2.27(c) shows the

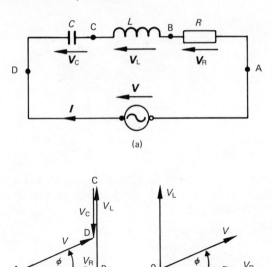

Fig. 2.27   **(a) Circuit diagram (b) String phasor diagram (c) Polar phasor diagram, general series circuit**

conventional 'polar' diagram, where all phasors are seen to start from an arbitrary point or pole O. This diagram is obtained by drawing all phasors emerging from O in parallel with the appropriate phasors in the 'string' diagram. The polar diagram is useful in that it shows more readily the phase angles between the current and the various voltages or between any two voltages. Furthermore, it is only necessary to scale up the phasors by multiplying them by $\sqrt{2}$ to obtain maximum values and to insert 'ω' to change the polar diagram to a rotating line diagram, the projection of which on to the vertical axis will give appropriate sinewaves. The string diagram is the link between the circuit and the polar diagram which makes possible the correct drawing of phasors. Furthermore, it is easier to decide from it the kind of calculations to be performed in order to establish the required relations for the circuit.

It is immediately obvious from Fig. 2.27(b) that

$$V^2 = V_R^2 + (V_L - V_C)^2$$

and expressing voltages in terms of current and constants of the circuit elements;

$$V^2 = I^2 R^2 + (IX_L - IX_C)^2 = I^2(R^2 + (X_L - X_C)^2)$$

therefore

$$\frac{V}{I} = (R^2 + (X_L - X_C)^2)^{\frac{1}{2}}$$

Expression $(R^2 + (X_L - X_C)^2)^{\frac{1}{2}}$ is termed the impedance of the circuit and is denoted by the symbol Z

i.e.,
$$Z = (R^2 + (X_L - X_C)^2)^{\frac{1}{2}} \tag{2.31}$$

Again from Fig. 2.27(b),

$$\tan \phi = \frac{V_L - V_C}{V_R} = \frac{IX_L - IX_C}{IR} = \frac{X_L - X_C}{R}$$

therefore,

$$\phi = \tan^{-1}\left(\frac{X_L - X_C}{R}\right) \qquad (2.32)$$

Thus,

$$V = IZ \qquad (2.33)$$

and since $I = I\angle0°$ and $V = V\angle\phi°$, multiplication of current by an impedance gives the voltage and its angle changes by ϕ. Hence angle ϕ may be associated with impedance Z. Substitution into (2.33) gives,

$$V\angle\phi = IZ\angle\phi° = I\angle0° \times Z\angle\phi°.$$

Therefore $\phi°$ must be attached to Z, i.e.,

$$Z = Z\angle\phi°.$$

Again Z does not symbolise a sinewave.

  Z may be represented pictorially by an impedance triangle as shown in Fig. 2.28(a). The angle between R and Z is ϕ as verified by equation (2.32). It is to be noted that when X_c is greater than X_L the triangle is inverted as in Fig. 2.28(b). The negative sign is associated with capacitive reactance to fit in with the phasor diagram.

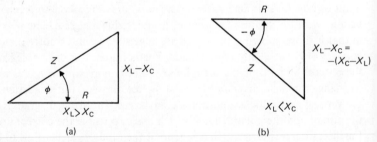

(a)   (b)

Fig. 2.28   Impedance triangles

  Impedance Z, therefore, is regarded as a constant of the circuit which determines the ratio of V to I and the phase angle between them, exactly as the reactance determines the ratio of V to I and stipulates a 90 degree phase angle between them in a circuit containing either pure inductance or pure capacitance.

QUESTION 5. A resistor of 34·6 Ω, a pure inductance of 0·255 H and a capacitor of 58·4 μF are connected in series across 200-V 50-Hz supply. Determine:

   1. The impedance of the circuit.
   2. The current in the circuit.
   3. Phase angle between supply voltage and the current.

  If the current is of the form $i = \sqrt{2}\, I \sin(2\pi ft)$ write the equation for the supply voltage in the same manner.

ANSWER. Impedance $= Z = \left(34 \cdot 6^2 + \left(2\pi 50 \times 0 \cdot 255 - \dfrac{10^6}{2\pi 50 \times 58 \cdot 4} \right)^2 \right)^{\frac{1}{2}}$

$$= (1200 + 400)^{\frac{1}{2}} = 40 \ \Omega$$

Phase angle ϕ equals $\tan^{-1} \dfrac{20}{34 \cdot 6} = 30°6'$, voltage leading the current since the inductive reactance is greater than the capacitive reactance.

Current $\qquad\qquad I = \dfrac{V}{Z} = \dfrac{200}{40} = 5$ A

and $\qquad\qquad i = \sqrt{2} \times 5 \sin (2\pi 50 t) = 7 \cdot 07 \sin 314t$ A

therefore $\qquad v = \sqrt{2} \times 200 \sin (2\pi 50 t + 30°6') = 282 \cdot 4 \sin \left(314t + \dfrac{\pi}{6} \right)$ V.

2.19 Series circuit containing more than three elements

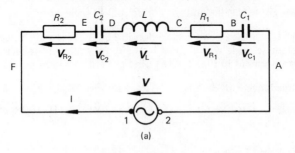

(a)

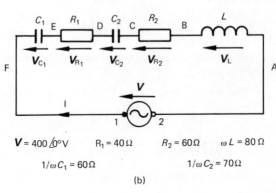

$\mathbf{V} = 400 \,\underline{/0°}\,$V $\qquad R_1 = 40 \,\Omega \qquad R_2 = 60 \,\Omega \qquad \omega L = 80 \,\Omega$

$1/\omega C_1 = 60 \,\Omega \qquad\qquad\qquad 1/\omega C_2 = 70 \,\Omega$

(b)

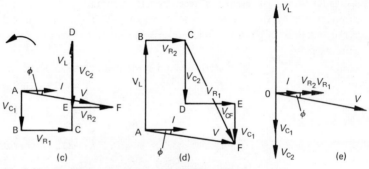

(c) (d) (e)

Fig. 2.29 Complex series circuit and its phasor diagrams

Consider a circuit made up of five elements as shown in Fig. 2.29(a). In Fig. 2.29(b) the same elements are again in series across the same supply but the order in which they are connected is different. In Fig. 2.29(c) and (d) voltage 'string' phasor diagrams are drawn for the circuits in Fig. 2.29(a) and (b) respectively. A current phasor is again used as a reference since it is common to all elements of the circuit.

Diagrams are drawn as before, starting at point A and setting phasors for resistance p.d. in parallel with the current phasor, inductive reactance p.d. at right angles to it and upwards, and capacitive reactance p.d. at right angles to it and downwards. It is immediately obvious that the string phasor diagram changes with the physical rearrangement of the circuit, although the current and phase relationship between it and the supply voltage is unaffected. If a polar diagram is drawn from each of the string diagrams, Fig. 2.29(e) is obtained in each case. If the voltage between any two points is required it is only necessary to join them on the string diagram and the resulting phasor gives its value and the correct phase angle.

QUESTION 6. Given the circuit arranged as in Fig. 2.29(a) and (b), determine the current and its phase angle with respect to the supply voltage for each case. Using the physical arrangement in Fig. 2.29(b), find the p.d. between points 'C' and 'F'.

ANSWER. Impedance Z in each case $= ((R_1 + R_2)^2 + (X_L - X_{C_1} - X_{C_2})^2)^{\frac{1}{2}}$

$$= ((40 + 60)^2 + (80 - 60 - 70)^2)^{\frac{1}{2}} = (10\ 100 + 2500)^{\frac{1}{2}}$$
$$= \sqrt{12\ 500} \triangleq 112\ \Omega$$

Therefore the current I in each case $= \dfrac{400}{112} = 3 \cdot 57$ A.

and
$$\phi = \tan^{-1} \left(\frac{50}{100} \right) = \tan^{-1}\ (0 \cdot 5)\ = 26 \cdot 6°. \ I \text{ leading } V.$$

The p.d. between C and F is obtained from the string diagram drawn to scale (Fig. 2.29(d)), or

$$V_{CF} = (143^2 + (250 + 214 \cdot 5)^2)^{\frac{1}{2}} = 530 \text{ V}$$

and
$$\phi = \tan^{-1} \left(\frac{250 + 214 \cdot 5}{143} \right) = \tan^{-1}\ (3 \cdot 93) = 68 \cdot 5°$$

V_{CF} lagging behind the current.

2.20   Parallel circuit containing R, L and C

Consider three basic elements: resistance R, inductance L and capacitance C connected in parallel across an a.c. supply, $v = \sqrt{2}\ V \sin \omega t$. Figure 2.30 shows this arrangement with phasor symbols inscribed next to the sense arrows for voltages and currents.

$$V = V \angle 0° \qquad \text{and symbolises } v\ = \sqrt{2}\ V \sin \omega t$$

$$I_R = I \angle 0° \qquad ,, \qquad ,, \qquad i_R = \sqrt{2}\ I_R \sin \omega t$$

$$I_L = I_L \angle -90° \ ,, \qquad ,, \qquad i_L = \sqrt{2}\ I_L \sin \left(\omega t - \frac{\pi}{2} \right)$$

$$I_C = I_C \angle 90° \qquad ,, \qquad ,, \qquad i_C = \sqrt{2}\ I_C \sin \left(\omega t + \frac{\pi}{2} \right)$$

$$I_T = I_T \angle \phi \qquad ,, \qquad ,, \qquad i_T = \sqrt{2}\ I_T \sin (\omega t + \phi)$$

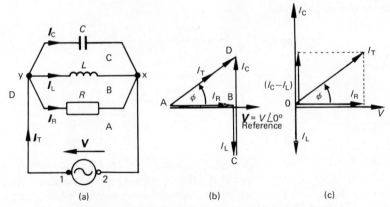

Fig. 2.30 **(a) Circuit diagram (b) Current 'string' phasor diagram (c) 'Polar' phasor diagram**

In order that the current 'string' phasor diagram may be constructed, 'Bows' notation used in mechanics may be adapted to the a.c. circuit as follows: All loops formed either by the a.c. source and one of the elements or by two elements are identified with capital letters A, B, C and D. Letter D denotes an 'outside' loop or space. Hence current flowing in resistor R is defined by spaces A and B and is marked on the phasor diagram as A → B.

Phasor V is used as a reference, and starting with point A currents I_R, I_L and I_C are drawn at the correct phase angles to the voltage. The string diagram constructed in Fig. 2.30(b) graphically represents an equation:

$$I_T = I_R + I_L + I_C \tag{2.34}$$

which is obtained by applying Kirchhoff's First Law to either junction x or y. The 'polar' phasor diagram in Fig. 2.30(c) is obtained by drawing all phasors emerging from a common pole O.

From the current 'string' phasor diagram in Fig. 2.30 it is seen that

$$I_T^2 = I_R^2 + (I_C - I_L)^2$$

and expressing the currents in terms of voltage and constants of the circuit elements

$$I_T^2 = \frac{V^2}{R^2} + \left(\frac{V}{X_C} - \frac{V}{X_L} \right)^2 .$$

is obtained.

Therefore
$$\frac{I_T}{V} = \left(\frac{1}{R^2} + \left(\frac{1}{X_C} - \frac{1}{X_L} \right)^2 \right)^{\frac{1}{2}} .$$

The ratio of I_T/V is termed the admittance of the circuit denoted by the symbol Y and expressed in Siemens, hence

$$Y = \left(\frac{1}{R^2} + \left(\frac{1}{X_C} - \frac{1}{X_L} \right)^2 \right)^{\frac{1}{2}} . \tag{2.35}$$

Again from Fig. 2.30(b)

$$\tan \phi = \frac{I_C - I_L}{I_R} = \frac{(V/X_C) - (V/X_L)}{V/R} = \frac{\omega C - (1/\omega L)}{1/R}$$

therefore,

$$\phi = \tan^{-1} \left(\frac{\omega C - 1/\omega L}{1/R} \right) \tag{2.36}$$

Thus

$$I = YV \tag{2.37}$$

and since $V = V \angle 0°$ and $I = I \angle \phi°$ multiplication of voltage by an admittance gives the current and its initial angle becomes $\phi°$. Hence angle ϕ may be associated with admittance Y, i.e. $Y = Y \angle \phi$. Pictorially this may be represented by an admittance triangle as shown in Fig. 2.31. G is a symbol denoting the conductance and B the susceptance of the circuit, both measured in Siemens. The negative sign in this case is associated with the reciprocal of the inductive reactance to agree with the phasor diagram in Fig. 2.30(b). It must be stressed that the relationship between G and R, B and X_C and X_L shown in Fig. 2.31 as well as equations (2.35) and (2.36) apply *only* to the simple circuit under discussion.

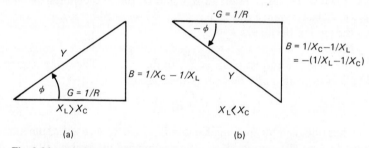

(a)   (b)

Fig. 2.31   Admittance triangles for pure parallel circuit

2.21   Parallel circuit-branches with more than one simple element

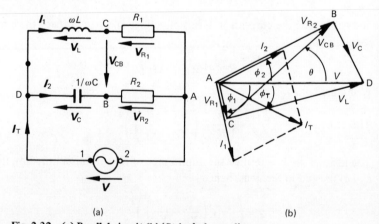

(a)   (b)

Fig. 2.32   (a) Parallel circuit (b) 'String' phasor diagram

Consider a circuit shown in Fig. 2.32. From the supply point of view the circuit is composed of two parallel branches. Hence the supply voltage V is used as a reference phasor acting from A to D. Next, current phasors are drawn, I_1 lagging the supply voltage by an angle ϕ_1 (inductive branch) and I_2 leading it by ϕ_2 (capacitive branch). Each of the two currents serves as a reference for the branch in which it flows.

Thus I_1 is a reference phasor for branch ACD and V_R and V_L are drawn in phase and $90°$ ahead of it respectively, forming the voltage string diagram ACD. Similarly V_R and V_C are in phase and $90°$ behind I_2 respectively forming the ABD portion of the diagram. If a voltmeter is connected between B and C, its reading will be numerically equal to the phasor CB. The phase angle of this voltage is either θ (Fig. 2.32(b)) with respect to the supply voltage, or $(180 - \theta)$ lagging behind it, depending entirely on the sense arrow direction placed on the circuit diagram. In Fig. 2.32(a) sense arrow points towards point B as being positive and therefore the phasor is pointing from C to B giving a phase angle of θ, i.e.,

$$V_{CB} = V_{CB}\angle\theta \text{ and symbolises } v_{CB} = \sqrt{2}\, V_{CB} \sin{(\omega t + \theta)}.$$

QUESTION 7. Given the circuit in Fig. 2.32(a) calculate:

1. Branch currents
2. Current taken from the supply.
3. The p.d. between points C and B assuming the reference condition shown by the sense arrow directed from C to B. $R_1 = 30\,\Omega, R_2 = 40\,\Omega$
$$\omega L = 40\,\Omega, 1/\omega C = 20\,\Omega,$$

the supply voltage $V = 200\angle0°$ (V).

ANSWER. $I_1 = \dfrac{200}{(30^2 + 40^2)^{\frac{1}{2}}} = \dfrac{200}{50} = 4 \text{ A}, \phi = \tan^{-1}\left(\dfrac{40}{30}\right) = 53°8' \text{ lagging } V$

$I_2 = \dfrac{200}{(40^2 + 20^2)^{\frac{1}{2}}} = \dfrac{200}{44\cdot7} = 4\cdot47 \text{ A}, \phi_2 = \tan^{-1}\dfrac{20}{40} = 26°31' \text{ leading } V.$

The phasor diagram is drawn to scale below (Fig. 2.33), from which

$$I_T = I_1 + I_2 = 6\cdot51\angle-10°36' \text{ A is obtained.}$$

The supply voltage (reference) is also drawn to scale and V_L, V_{R_1}, V_C and V_{R_2} are drawn as described previously. (Angles at B and C are equal to $90°$.)

From the diagram:	By calculations:
$V_{R_1} = 120$ V	$V_{R_1} = 30 \times 4 = 120\angle-53°8'$ V
$V_L = 160$ V	$V_L = 40 \times 4 = 160\angle30°52'$ V
$V_{R_2} = 179$ V	$V_{R_2} = 40 \times 4\cdot47 = 178\cdot8\angle26°31'$ V
$V_C = 89\cdot5$ V	$V_C = 20 \times 4\cdot47 = 89\cdot4\angle-63°29'$ V
$V_{cb} = 200\angle63°40'$ V	

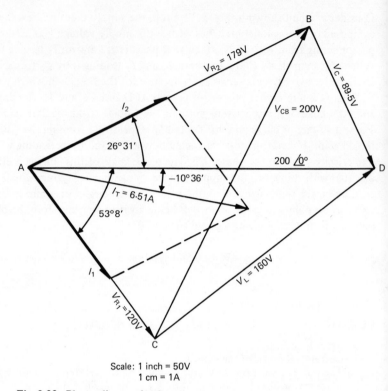

Scale: 1 inch = 50V
1 cm = 1A

Fig. 2.33 Phasor diagram for Question 7

2.22   Power in single phase a.c. circuits

GENERAL EXPRESSION FOR POWER AND POWER FACTOR

The instantaneous power $p = vi$ in an a.c. circuit varies continuously with time. It is therefore the net or average power transfer between a source and the load which is important.

If P is the average power transferred in the circuit where V is the effective (r.m.s.) voltage and I is the effective (r.m.s.) value of the current, then it is found that P is often, but not always, less than the product VI. Hence VI must be multiplied by a constant, which may have any value between 0 and 1. The constant is called a 'power factor' (p.f.) and the general expression for an average power becomes:

$$P = (\text{p.f.}) \; VI \tag{2.38}$$

where (p.f.) is the power factor and VI is the total volt-ampere product measured in (VA).

From equation (2.38) the power factor can be defined as follows:

$$(\text{p.f.}) = \frac{P}{VI} = \frac{\text{average power (watts)}}{\text{r.m.s. voltage (volts) x r.m.s. current (amperes)}} \tag{2.39}$$

The (p.f.) being a ratio has no unit. The equations (2.38) and (2.39) are general and apply to a.c. circuits whatever the shape of the voltage and current waveforms.

POWER AND POWER FACTOR FOR SINUSOIDAL EXCITATION

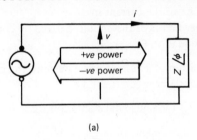

(a)

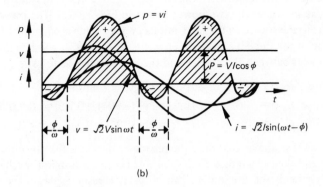

(b)

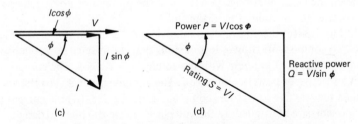

(c)   (d)

Fig. 2.34 Appertaining to power in sinusoidal a.c. circuit (a) circuit
(b) waveforms of v, i and p (c) phasor diagram (d) power
triangle

An a.c. circuit consisting of a source and the load of impedance Z is shown in Fig.
2.34(a). If the potential difference $v = \sqrt{2}V \sin \omega t$ and the current $i = \sqrt{2}I \sin$
$(\omega t - \phi)$, the latter lagging the p.d. by an angle ϕ, then the instantaneous power
$p = vi = \sqrt{2}V \sin \omega t \times \sqrt{2}I \sin (\omega t - \phi) = 2VI \sin \omega t \sin (\omega t - \phi) = 2VI/2 \{\cos \phi -$
$\cos (2\omega t - \phi)\} = VI \cos \phi - VI \cos (\omega t - \phi)$.

The term $VI \cos \phi$ is independent of time t and therefore its value is constant. The
term $VI \cos (2\omega t - \phi)$ varies with time at twice the angular frequency (2ω) of the
voltage. Since the average of any cosine wave over the full cycle is zero, therefore the
average value of the second term is zero. Thus the average power transfer from the
source to the load is equal to the first term only, i.e.,

$$P = VI \cos \phi \qquad (2.40)$$

where V is the r.m.s. value of the voltage in
      I is the r.m.s. value of the current in
and $\cos \phi$ is the power factor of the circuit, derived by comparison of the equation
      (2.40) with the equation (2.38).

Thus for sinusoidal variation of a voltage and a current, the power factor is always
equal to a cosine of an angle between them, i.e.,

$$\text{(p.f.) for sinusoidal excitation} = \cos \phi \qquad\qquad (2.41)$$

The waveforms of v, i and p are shown in Fig. 2.34(b).

When the power curve p lies above the time axis, its value is positive and the source
transfers power to the load. When the curve is negative, the load returns the power to
the source, as shown by the two power arrows in Fig. 2.34(a). The transfer however
is greater from the source to the load and its average value is equal to $VI \cos \phi$ watts.

Referring to the three basic circuit elements R, L and C it will be remembered that
the phase angles between the voltage and the current were as follows:

for resistance $R, \phi = 0,$ $\therefore$ (p.f.) $= \cos 0° = 1$ and so $P = VI$
for inductance $L, \phi = -90°,$ $\therefore$ (p.f.) $= \cos (-90°) = 0$ and so $P = 0$
for capacitance $C, \phi = 90°,$ $\therefore$ (p.f.) $= \cos (90°) = 0$ and so $P = 0$

It is clear from the above, that only the resistance develops the power, whereas the
inductance and the capacitance are responsible for the continuous exchange of energy
between their fields and the source. This property is responsible for the discrepancy
between the value of average power P and the product of r.m.s. values VI and
necessitates the introduction of the (p.f.).

In Fig. 2.34(c) the phasor diagram of V and I is drawn. Resolving the current phasor
into two components shown, it is obvious that the average power depends on the
current component *in phase* with the voltage, i.e., $I \cos \phi$. The component is referred
to as an 'active or power component'. The component at $90°$ to the voltage V is
termed the reactive or wattless component. In Fig. 2.34(d) the current phasor
components are multiplied by the voltage V, giving the power triangle. The three sides
of the triangle represent:

1. The average power $P = VI \cos \phi$
2. The reactive power (reactive volt-amperes) $Q = VI \sin \phi$
3. The total volt-amperes (rating) $S = VI$

For larger devices the three quantities are usually multiplied by a factor of 1000 and
thus given in kW, kVAr and kVA.

THREE-PHASE ALTERNATING CURRENT CIRCUIT THEORY

2.23   Three-phase alternating current

Three single phase sources of alternating current connected together in a particular
manner result in a single 'three-phase' alternating current system. The sources are
identical in every respect, except that they change their polarity at different intervals

and to provide symmetry their current waveforms are $120°$ out of phase with each other.

The first part of the theory deals with basic ways of connecting the sources to form a single system. The connections described are:

(i) Mesh or delta (Δ)
(ii) Star or Wye (Y)
(iii) Open delta or Vee (V)

The second part of the theory describes various ways of loading the three phase supply systems obtained.

The systems have many advantages over a single phase system, such as saving in a number of connections between supply and the load (lower cost of materials), transfer of electrical energy without fluctuations, and more efficient use of space inside the electrical machines. For these reasons, as well as others, practically all electrical power generated nowadays in the world is of a three phase nature.

2.24 Three single-phase sources of e.m.f.

Three separate e.m.f. sources produce a sinusoidal potential difference across their terminals of equal frequency and r.m.s. value but their initial angles differ by $120°$.

Let the terminals of the first source be designated by letters R and R', its reference polarity be from R' to R and its potential difference $v_R = \sqrt{2}\,V \sin(\omega t)$ volts, symbolised by $V_R\angle0°$. Similarly the other two sources are characterised by equations and symbols as follows:

second source $\qquad v_Y = \sqrt{2}\,V \sin(\omega t - 120°), V_Y\angle{-120°}$, and
third source $\qquad v_B = \sqrt{2}\,V \sin(\omega t - 240°), V_B\angle{-240°}$

Figure 2.35 shows the sources with their arrows, the time plots of their potential differences and their phasor diagrams.

Connecting the sources in series by joining terminals R' to Y and Y' to B as in Fig. 2.36(a), the instantaneous potential difference between B' and R is given by an application of Kirchhoff's voltage law.

$$v_{B'R} = v_R + v_Y + v_B$$

therefore
$$v_{B'R} = \sqrt{2}\,V \sin \omega t + \sqrt{2}\,V \sin(\omega t - 120°) + \sqrt{2}\,V \sin(\omega t - 240°)$$
$$= \sqrt{2}\,V \{\sin \omega t + \sin \omega t \cos 120° - \cos \omega t \sin 120°$$
$$+ \sin \omega t \cos 240° - \cos \omega t \sin 240°\} = \sqrt{2}\,V \{\sin \omega t$$
$$-0{\cdot}5 \sin \omega t + 0{\cdot}866 \cos \omega t - 0{\cdot}5 \sin \omega t - 0{\cdot}866 \cos \omega t\}$$
$$= 0 \text{ volts}$$

The same result is obtained graphically. The diagram in Fig. 2.36(a) is relabelled with phasor symbols, shown in brackets, from which voltage string diagram is drawn starting from point B. Each phasor is drawn at the correct angle to the horizontal and the triangle in Fig. 2.36(b) is obtained. The resultant phasor is zero, since the points R and B' are coincident.

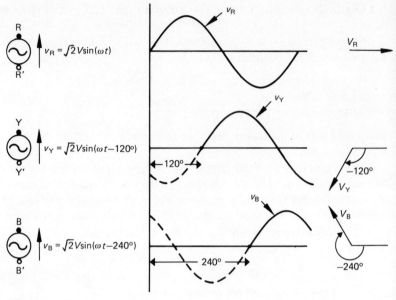

Fig. 2.35   Three single-phase sources

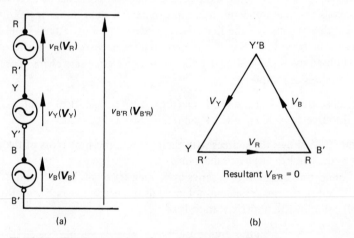

Fig. 2.36   Three sources in series and their phasor diagrams

This result suggests that if points R and B′ are joined together there will be no flow of current through the three sources. Thus the delta may be closed without loading the sources.

2.25   Mesh or delta connection

The three e.m.f. sources are now arranged in the form of a triangle, Fig. 2.37(a), each supplying a load of impedance $ZL\phi$. Since each sources' potential difference is numerically the same i.e., V volts, hence each delivers an equal current of I amperes

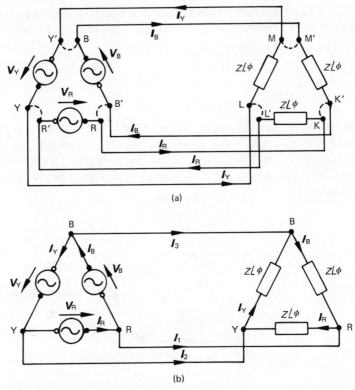

Fig. 2.37 Development of delta connection

to the load. The phase angles however, differ because of the $120°$ shift between their
e.m.f.'s.

Thus symbolically the currents are represented by:

$$I_R = IL{-}\phi, I_Y = IL{-}120° - \phi, I_B = IL{-}240° - \phi$$

Note that negative sign in front of ϕ is necessary because impedance is assumed to be
inductive and thus causes the currents to lag behind the voltages producing them. If
terminals R and B', B and Y', Y and R' are joined together, (dotted line in Fig. 2.37(a))
the current distribution round the three sources which now form a closed loop is
unaffected, because by the conclusions of the previous paragraph the resultant of
$V_R + V_Y + V_B = 0$.

Three sources thus connected are said to be in delta or mesh connection. Furthermore, if the impedance of the leads joining the sources to the loads is negligible, as is
usually the case for short distances, then voltage drops along them are negligible and
points K and K' are at the same potential as point R, L and L' as Y, and M and M' as B.
Hence 'corners' of the load network can also be connected together without affecting
current distribution round the load delta. Each pair of the conductors may now be
replaced by one lead, which will carry the difference of the currents in the original
pair. The final network then is shown in Fig. 2.37(b).

The positive direction of the currents in the leads (known as line currents) is assumed to be from the supply towards the loads. They are designated I_1, I_2 and I_3. Hence:

$$I_1 = I_R - I_B$$
$$I_2 = I_Y - I_R$$
$$I_3 = I_B - I_Y$$

The above equations are obtained, either from Fig. 2.37(a) or by applying Kirchhoff's First Law to points R, Y, and B in Fig. 2.37(b). The solution is obtained graphically and is shown in Fig. 2.38(a) where string phasor diagram for potential differences is drawn, with currents I_R, I_Y and I_B lagging by an angle ϕ behind their respective voltages.

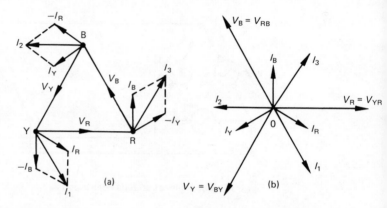

Fig. 2.38   String and polar phasor diagram for delta connection

I_1 which is the difference between I_R and I_B is obtained by reversing phasor I_B, transferring it to point Y of the voltage triagle and adding it to I_R by the usual parallelogram law. I_2 and I_3 are obtained in a similar manner. Polar diagram is shown in Fig. 2.38(b) in which all phasors are seen to emerge from an arbitrary pole O and are drawn in parallel with the appropriate phasors in the string diagram. This diagram gives phase angles between various phasors at a glance. For example the angle between I_R and I_1 is 30° with the latter lagging behind the former since anticlockwise rotation is assumed as standard for phasor representation.

The magnitude of the line current I_1 is obtained by remembering that I_R and I_B are numerically equal and therefore I_R, I_B and I_1 form equilateral triangle (Fig. 2.38(a)):

Hence $I_1 = I_R \cos 30° + I_B \cos 30°$

and putting $I_R = I_B = I_{phase}$,

$$I_1 = I_{ph} \cos 30° \times 2 = 2 I_{ph} \frac{\sqrt{3}}{2} = \sqrt{3} I_{ph} \text{ i.e.,}$$

$$I_{line} = \sqrt{3} I_{phase} \tag{2.42}$$

Furthermore, from Fig. 2.37(b) it is obvious that the potential difference between any two lines is equal to the potential difference of each e.m.f. source, i.e.,

$$V_{\text{phase}} = V_{\text{line}} \tag{2.43}$$

2.26 Phase sequence

The phase sequence R–Y–B is regarded as STANDARD. In the previous paragraph delta connection was formed by connecting terminals R' to Y, Y' to B and B' to R of the three separate e.m.f. sources with a STANDARD phase sequence. This connection reproduced in Fig. 2.39(a) is shown supplying three lines labelled 1, 2 and 3. The potential differences between the lines are designated:

$V_{2,1}$ between lines 2 and 1

$V_{3,2}$ between lines 3 and 2

$V_{1,3}$ between lines 1 and 3,

their reference direction being given by the sense arrows as well as the order of the suffixes placed next to the symbol V. From this circuit diagram the string phasor diagram is drawn in Fig. 2.39(b), starting with point B, each phasor being drawn at the correct angle to the horizontal. In turn the polar diagram is obtained by drawing all phasors emerging from a common pole O and in parallel with their counter-parts in the string diagram. From the latter it is seen that the sequence of line voltages in an anticlockwise direction of rotation is:

$V_{2,1}$ followed by $V_{3,2}$ followed by $V_{1,3}$.

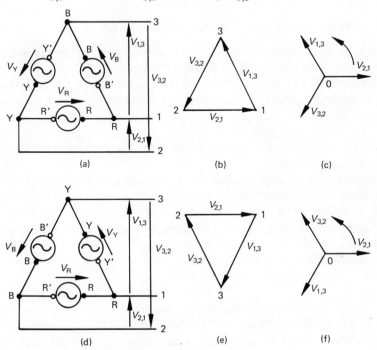

Fig. 2.39 Appertaining to phase sequence

Let the connection of the delta be modified by interchanging the positions of the two sources marked YY$'$ and BB$'$. The resulting diagram of connection is given in Fig. 2.39(d). From it the string phasor diagram and the polar diagram are drawn in Fig. 2.39(e) and Fig. 2.39(f) respectively.

On examining the polar diagram the sequence of the line voltages in an anticlockwise direction is found to be:

$$V_{2,1} \text{ followed by } V_{1,3} \text{ followed by } V_{3,2}.$$

In the clockwise direction however, the sequence is:

$$V_{2,1} \text{ followed by } V_{3,2} \text{ followed by } V_{1,3},$$

which is the same as in the first case.

It is seen therefore, that an interchange of the position of any two sources is equivalent to the change in the *direction* of rotation of phasors representing line voltages.

The first case is known as the ANTICLOCKWISE sequence, whereas the second case is known as the CLOCKWISE sequence. The same result may be obtained by keeping the sources connected as in the first case but interchanging line connection. For example: connecting line 2 to the junction of Y$'$B and line 3 to the junction of YR$'$.

Note that when the direction of phasor rotation is considered the observer is stationary and the phasors rotate.

2.27   Delta connection with one phase reversed

Consider the delta connection of the three sources obtained by joining terminals as follows: Y to R$'$, R to B, and B$'$ to Y$'$. Before however, the last two points are joined together, the potential difference between them is investigated. Figure 2.40(a) shows delta connection as described above. Applying Kirchhoff's Second Law equation:

$$V_{Y'B} = V_R + V_Y - V_B$$

is obtained, which gives p.d. between Y$'$ and B$'$ for the reference direction indicated by the sense arrow.

The phasor string voltage diagram in Fig. 2.40(b) which is a graphical representation of the equation, is drawn starting from point Y$'$. Phasor V_Y is drawn at an angle of $-120°$ to the horizontal, V_R at $0°$ and V_B at $-60°$. The phasor V_B is shown $180°$ out of phase with its normal position because of the negative sign in front of it. The voltage $V_{Y'B'}$ is obtained by completing the string. From the geometry of the figure it is obvious that the value of the $V_{Y'B'}$ is twice that of the phase voltage, i.e.,

$$V_{Y'B'} = 2\,V_{ph} \qquad \qquad (2.44)$$

and it lags $60°$ behind the V_R. The phase angles between the various phasors are more easily seen from the polar diagram shown in Fig. 2.40(c). When the two points are joined together, voltage $V_{Y'B'}$ produces a circulating current round the delta limited only by the impedances of the three sources.

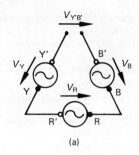

(a)

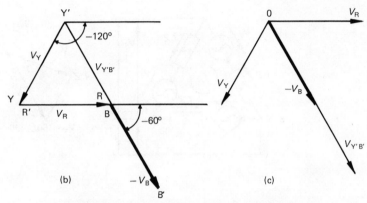

(b)   (c)

Fig. 2.40   (a) Circuit diagram (b) String voltage diagram (c) Polar phasor diagram, incorrect delta

Equation (2.44) is useful in determining the correct connection of the delta, if reference directions of the sources are unknown, or their markings not clear. It is only necessary to place the voltmeter across the 'last' corner of the delta, before joining it, and if its reading is zero, the connection is correct, If on the other hand it reads twice the phase voltage, then one of the sources is reversed. Reversing each source in turn until the voltmeter reads zero enables correct connection of the delta to be arrived at.

QUESTION 8. Potential differences of the three sources are given by:

$$V_R = 150\angle 0° \text{ V}, \; V_Y = 150\angle -120° \text{ V and } V_B = 150\angle -240° \text{ V}$$

The sources are connected in delta with the third source inadvertently reversed. Calculate the circulating current round the delta, if the internal resistance and inductive reactance of each source are 3 Ω and 4 Ω respectively. Find also the potential differences between the three lines of such a delta network.

ANSWER. I circulating $= \dfrac{2 V_{ph}}{Z_R + Z_Y + Z_B} = \dfrac{2 \times 150}{3\sqrt{3^2 + 4^2}}$

$$= \frac{300}{15} = 20 \text{ A lagging behind}$$

$$V_{Y'B'} \text{ by } \phi = \tan^{-1}\left(\frac{4}{3}\right) = 53°10'.$$

The circuit diagram is shown below in which each source is represented by its linear source equivalent i.e., constant e.m.f. of 150 V in series with its internal impedance of 5 Ω. Assuming the same reference direction of the voltage $V_{Y'B'}$ as in Fig. 2.40(a), the circulating current I_c flows in an anticlockwise direction round the delta as shown. The equations to find the line voltages are obtained by applying Kirchhoff's Second Law to each loop, i.e., starting from the point B in an anticlockwise direction through point Y and back to point B via the line voltage we have:

$$V_Y - I_c Z - V_{BY} = 0, \text{ therefore } V_{BY} = V_Y - I_c Z$$

Similarly

$$V_{YR} = V_R - I_c Z$$

and

$$V_{RB} = -(V_B + I_c Z)$$

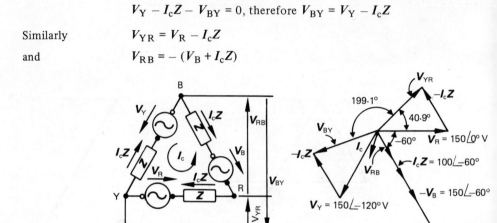

Fig. 2.41 Appertaining to Question 8

$$V_{Y'B'} = 300\angle -60° \text{ V}$$

Drawing the polar phasor diagram to scale the line voltages are obtained by consistent application of these equations, i.e., to find V_{BY}, $I_c Z = 20 \times 5 = 100$ V is reversed and added to the phasor V_Y; to find V_{YR} again $I_c Z$ is reversed and added to V_R; to find V_{RB}, $I_c Z$ is added to V_B and the result reversed.

Thus

$$V_{YR} = 132 \cdot 5\angle 40 \cdot 9° \text{ V}$$
$$V_{BY} = 132 \cdot 5\angle 199 \cdot 1° \text{ V}$$
$$V_{RB} = 50\angle -60° \text{ V}$$

This example shows that the incorrect delta connection does not produce a balanced system of line voltages.

2.28 Star or Y connection

In Fig. 2.42(a) the same three e.m.f. sources which were used to form delta connection are arranged physically to suggest Y configuration. Each source is still depicted supplying its own load of impedance $Z\angle\phi$. The three resulting currents, each lagging by an angle ϕ behind the voltage producing it, are given by the equations:

$$i_R = \sqrt{2} I \sin(\omega t - \phi)$$
$$i_Y = \sqrt{2} I \sin(\omega t - 120° - \phi)$$
$$i_B = \sqrt{2} I \sin(\omega t - 240° - \phi)$$

where $I = \dfrac{V}{Z}$

The sense arrows, indicating reference directions for the potential differences and the above currents are included in the figure according to the convention of Chapter 1.

Examination of the arrangement suggests that the three conductors which run together may be replaced by a single conductor if points R', Y' and B' are joined together to form a 'star' point on the supply side, and a similar point on the load side of the circuit. The three single conductors may now be replaced by one termed a neutral which carries the sum of the three currents, i.e.,

$$i_N = i_R + i_Y + i_B$$

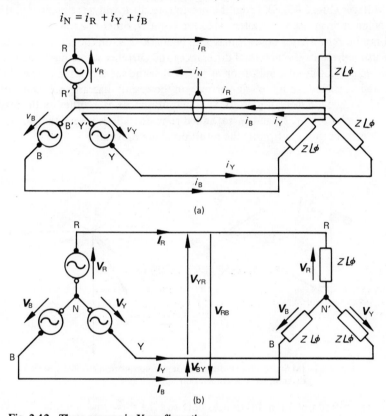

(a)

(b)

Fig. 2.42 Three sources in Y configuration

This equation is a consequence of Kirchhoff's First Law applied to either of the two star points.

Therefore,

$$i_N = \sqrt{2}\,I \sin{(\omega t - \phi)} + \sqrt{2}\,I \sin{(\omega t - \phi - 120°)} +$$
$$\sqrt{2}\,I \sin{(\omega t - \phi - 240°)} = \sqrt{2}\,I\{\sin{(\omega t - \phi)} +$$
$$\sin{(\omega t - \phi)} \cos 120° - \sin 120° \cos{(\omega t - \phi)} +$$
$$\sin{(\omega t - \phi)} \cos 240° - \sin 240° \cos{(\omega t - \phi)}\}$$
$$= \sqrt{2}\,I\{\sin{(\omega t - \phi)} - 0{\cdot}5 \sin{(\omega t - \phi)} - 0{\cdot}866 \cos{(\omega t - \phi)}$$
$$- 0{\cdot}5 \sin{(\omega t - \phi)} + 0{\cdot}866 \cos{(\omega t - \phi)}\} = 0.$$

The above result shows that since the three currents add up to zero, the neutral wire is unnecessary and can be omitted. It must be noted however that this is only true when the impedances of the three loads are equal in magnitude and phase angle. The new circuit arrangement known as star or Y connection is shown in Fig. 2.42(b).

Instead of the instantaneous values, the phasor symbols are now shown next to the sense arrows, so that the string phasor diagram can be drawn. Starting from point N (neutral), V_R is drawn horizontally to point R, V_Y is drawn from N to point Y at 120° behind V_R and V_B from N to B at 120° behind V_Y.

The currents I_R, I_Y and I_B are drawn lagging by an angle ϕ behind their respective phase voltages (Fig. 2.43(a)). From the previous result it will be remembered that point N′ must be at the same potential as the point N, otherwise the neutral current would not be zero. Hence point N may also be marked N′ on the phasor diagram, which also represents the potential differences and currents across the three loads. As mentioned previously, the indication as to whether the supply or the load end is being considered, is gained from the sense arrows on the circuit diagram (p.d. and current sense arrows agree in direction on the supply, but oppose each other on the load element). If phasors I_R, I_Y and I_B are added geometrically it is seen that their resultant is zero, thus confirming the result obtained previously.

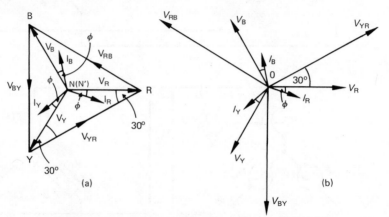

Fig. 2.43   (a) String phasor diagram for star connection (b) Polar phasor diagram for star connection

In order to find potential differences between the three line conductors, it is necessary first to decide the reference direction of the sense arrows. The choice is purely arbitrary, since whether p.d. of Red line being at higher potential than Yellow line is regarded as positive or vice versa, does not depend on the phase voltages nor the currents flowing in the lines. The only thing of importance is consistency; i.e., if p.d. of Red line with respect to Yellow is assumed as reference, then p.d. of Yellow line with respect to Blue and that of Blue with respect to Red must be taken. Putting arrows as just described in Fig. 2.42(b) the three line p.d.'s are obtained from the string phasor diagram in Fig. 2.43(a) by joining points Y to R, R to B and B to Y. The arrow heads are put on these lines in the same directions as the sense arrows on the circuit diagram.

In Fig. 2.43(b) polar diagram is shown, which is obtained from Fig. 2.43(a) by drawing all the phasors emerging from an arbitrary point O, that is unrelated to any point on the circuit diagram. It must be mentioned that if sense arrows in Fig. 2.42(b)

for line p.d.'s were reversed, the magnitude of the line p.d. phasors would not be affected, but they would be $180°$ out of phase with those shown in Fig. 2.43(b). Considering the portion of the string diagram (Fig. 2.43(a)) containing V_R, V_Y, and V_{YR}, the figure is a triangle, the angles of which have magnitudes shown. Furthermore, phase voltages are all equal in magnitude, i.e., $V_R = V_Y = V_B = V_{ph}$

Hence magnitude of $V_{YR} = 2\, V_{ph} \cos 30°$

$$= 2\, V_{ph}\, \frac{\sqrt{3}}{2} = \sqrt{3}\, V_{ph}.$$

But from the string diagram it is obvious that the magnitudes of all three line voltages are also equal, i.e. $V_{YR} = V_{BY} = V_{RB} = V_{line}$.
Hence for this connection

$$V_{line} = \sqrt{3}\, V_{phase} \tag{2.45}$$

and leads phase voltage by $30°$ (see polar diagram).

Again considering the circuit diagram it will be noticed that the currents flowing through the e.m.f. sources flow through the line conductors and through the load elements, hence phase current is one and the same as the line current, i.e.,

$$I_{line} = I_{phase} \tag{2.46}$$

both in magnitude and in phase.

In the star connection considered above, the three e.m.f. sources were so arranged that positive sequence of phase rotation was obtained i.e., V_R was followed by V_Y and V_Y was followed by V_B in an anticlockwise direction. This produced the following sequence for line voltages: V_{YR} followed by V_{BY} followed by V_{RB}.

As in delta the negative or clockwise phase sequence can be obtained for star connection, either by interchanging physical position of any two e.m.f. sources or by changing over two line conductors to the connection shown in Fig. 2.42(b).

2.29   Star connection with one phase reversed

In the previous paragraph the star connection was formed using three e.m.f. sources by connecting terminals R', Y' and B' together. If points R, Y and B were to be joined together to form a neutral point, the resulting system would be the same as before, except that the voltage phasor diagram would have to be drawn rotated through $180°$ in an anticlockwise direction (sense arrow directions for source e.m.f.'s are those in Fig. 2.42(a)).

Suppose however, that star point is obtained by joining together terminals R', Y and B'. This is equivalent to a reversal of e.m.f. source marked YY' (or a reversal RR' and BB' sources if the normal system is that in which the neutral point is a junction of R, Y and B terminals).

In Fig. 2.44(a) the circuit diagram of such a system is shown. Sense arrows indicating V_R, V_Y and V_B as well as line voltages are drawn exactly as for normal star connection, but since reference direction for V_Y is really from Y' to Y, the sense arrow directed from Y to Y' represents $-V_Y$. With this in mind the voltage string phasor diagram is drawn in Fig. 2.44(b) showing phase and line voltages. From it, it is

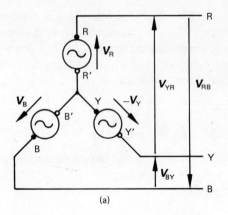

(a)

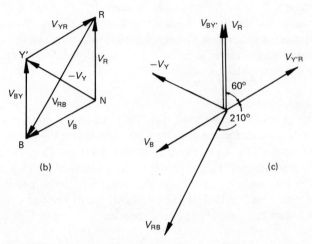

(b)   (c)

Fig. 2.44   Incorrect Y connection

seen that the line voltages are no longer balanced in magnitude nor in phase. Their
values are as follows:

$$V_{RB} = \sqrt{3}\, V_{phase} \text{ and lags } V_R \text{ by } 210°$$
$$V_{BY'} = V_{phase} \quad \text{ and is in phase with } V_R$$
$$V_{Y'R} = V_{phase} \quad \text{ and lags } V_R \text{ by } 60°$$

The phase angles enumerated above are clearly shown in the polar voltage diagram in
Fig. 2.44(c).

   The above result provides a practical check on whether star connection has been
properly formed or not. If the measurement of line voltages reveals that one reading
is $\sqrt{3}$ times that of the other two, then one phase is reversed. To find which it is, again
line voltages are measured until the maximum reading is obtained, the terminal which
is not connected to a voltmeter belongs to the phase which is reversed.

2.30   Open delta or V connection

If one e.m.f. source is removed from the delta connection the result is known as an open delta or V connection. The circuit in Fig. 2.45(a) illustrates just such a connection, from which an e.m.f. source marked RR′ is omitted.

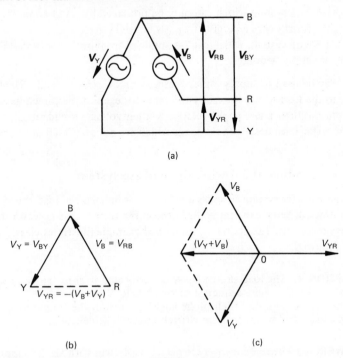

(a)

(b)   (c)

Fig. 2.45   Circuit and phasor diagrams for V connection

Starting from point R and drawing V_B and V_Y at the correct phase angle to the horizontal, the string diagram is obtained in Fig. 2.45(b). Examination of the sense arrows in the circuit diagram shows that $V_B = V_{RB}$ (voltage between R and B lines), $V_Y = V_{BY}$ (voltage between B and Y lines), whereas V_{YR} is the 'equilibrant' of the phase voltages V_B and V_Y, that is phasor joining points R and Y in the string diagram shown by a dotted line. Thus although there is no e.m.f. source between Y and R lines, nevertheless voltage between them is in every respect that of a normal delta connection. The removal, therefore, of one of the e.m.f. sources does not affect the line voltages of the system, i.e., three-phase balanced system is obtained as seen clearly from the polar diagram in Fig. 2.45(c). The current distribution is, however affected.

2.31   Three- and four-wire supply systems

Previous discussion was confined to the production of the three phase supply by interconnecting three single phase e.m.f. sources. In practice however, the sources take the form of three separate windings either built inside one casing, thus forming a modern three-phase synchronous generator, or wound on the common magnetic core of a three-phase transformer.

The two ways of interconnection discussed so far, in general produce two main systems:

(i) a 4-wire system which can only be produced by a *star*-connected winding with the fourth wire emerging from its neutral point;

(ii) a 3-wire system, which can be obtained from both star or delta configuration of supply windings.

Both may be used to supply either *balanced* or *unbalanced* loads. The former term refers to the load which is exactly the same for each of the three phases, i.e., the currents and their phase angles relative to their voltages are identical.

The latter term describes the loads which are not restricted in this way.

3.32   Loading of a 4-wire, three-phase system

Whether a 4-wire system supplies a balanced or unbalanced load the method of calculating currents is the same for both. In the former case however, the neutral wire carries no current. The following numerical example illustrates clearly the procedure to be followed.

QUESTION 9. The load shown below is connected to three-phase, 4-wire, 240-V, 50-Hz supply, the phase sequence of which is R, Y, B.

Find the currents and their phase angles in each of the four connections. The values of resistances and reactances are written next to each element.

ANSWER. (i) The sense arrows are placed as shown, with all the supply voltage arrows as well as those across the load elements emerging from the neutral wire. The line current arrows point towards, and a neutral current arrow away, from the load.

(ii) Since by Kirchhoff's voltage law

$$V_R = V_R', V_Y = V_Y' \text{ and } V_B = V_B'$$

Hence,
$$I_R = \frac{240}{120} = 2 \text{ A and is in phase with } V_R$$

$$I_Y = \frac{240}{80} = 3 \text{ A and lags } V_Y \text{ by } 90°$$

and
$$I_B = \frac{240}{(40^2 + 30^2)^{\frac{1}{2}}} = \frac{240}{50} = 4 \cdot 8 \text{ A and leads } V_B \text{ by } \phi = \tan^{-1}\left(\frac{30}{40}\right)$$
$$= 36 \cdot 9°$$

(iii) Applying Kirchhoff's current law to point N, the neutral current $I_N = I_R + I_Y + I_B$. This equation is solved graphically by drawing phasor diagram to scale and adding I_Y to I_B and then their sum to I_R, from which $I_N = 6 \cdot 1$ A and leads V_B by an angle of 25°.

(iv) Finally, the voltage V_B, is the sum of V_1 and V_2 each of which can be obtained either by calculation:

$$V_1 = 4 \cdot 8 \times 40 = 192 \text{ V in phase with } I_B$$
and
$$V_2 = 4 \cdot 8 \times 30 = 144 \text{ V lagging } I_B \text{ by } 90°,$$

or by drawing phasors to scale as shown.

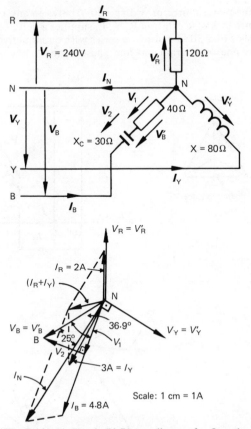

Fig. 2.46 (a) Circuit (b) Phasor diagram for Question 9

2.33 Loading of a 3-wire, three-phase system

Just as the supply windings may be connected in star or delta, so too loads may be arranged in either of the two configurations. When the load elements are joined in STAR and the load is *balanced*, then the solution is exactly the same as for the case of a 4-wire system. This is so because, although the fourth wire is absent, the neutral points of the supply and the load are at the same potential and each phase current is calculated by dividing the appropriate phase voltage by the impedance of each load.

When the loads are *unbalanced* the neutral points are no longer at the same potential and phase voltages across three elements are unequal. This is because the absence of a neutral wire does not allow the flow of unbalanced current (see Question 9) and causes voltage unbalance instead.

The solution of such a problem demands the use of complex algebra and cannot be obtained by drawing phasors to scale. This lies outside the scope of this book.

When the elements of the load are connected in DELTA, the solution can be obtained from the phasor diagram irrespective of whether the load is balanced or not. Furthermore, the method of calculating currents and voltages is exactly the same for both, as the following example shows:

QUESTION 10. A three-phase source providing 400 V between any two lines, at 50 Hz and having standard anticlockwise rotation R, Y, B, supplies a delta connected load, each leg of which consists of a 30 Ω resistor in series with an inductive reactance of 40 Ω. Draw a complete phasor diagram and find phase currents, line currents, and the phase angle for each.

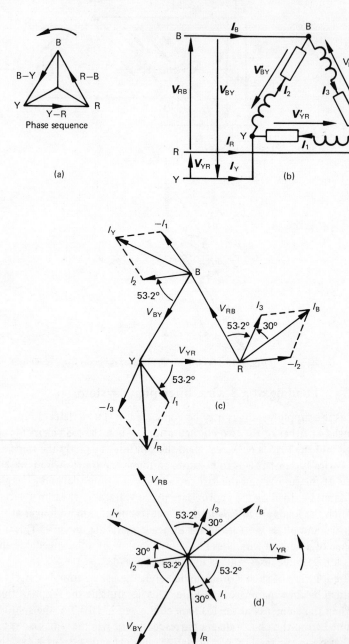

Fig. 2.47 Appertaining to Question 10

ANSWER. The circuit together with sense arrows indicating line voltages, line currents and phasor currents is shown above. In drawing it and placing the arrows the following points are helpful:

(i) Draw first the delta load and mark three corners R, Y and B, starting with the right-hand corner and moving the pen in a clockwise direction.

(ii) Connect three line conductors to the three corners as shown. Notice that lines R and Y are crossed over to fit in with the delta connected supply.

(iii) Place line voltage sense arrows exactly as above.

(iv) Place line current arrows pointing *always* towards the load.

(v) Place voltage arrows across each load branch in such a way that their tips oppose those of line voltages.

(vi) The current arrows in each branch can then be placed opposing those of phase voltages.

From the circuit diagram it is then clear that:

$$I_1 = \frac{V_{YR}}{(30^2 + 40^2)^{\frac{1}{2}}} = \frac{400}{50} = 8 \text{ A and lags } V_{YR} \text{ by } \phi = \tan^{-1} \frac{40}{30}$$
$$= 53 \cdot 2°$$
$$\text{i.e., } I_1 = 8\angle -53 \cdot 2°$$

$$I_2 = \frac{V_{BY}}{(30^2 + 40^2)^{\frac{1}{2}}} = 8 \text{ A and lags } V_{BY} \text{ by } \phi = 53 \cdot 2°$$
$$\text{i.e., } I_2 = 8\angle -120° - 53 \cdot 2° = 8\angle -173 \cdot 2°$$

$$I_3 = \frac{V_{BR}}{(30^2 + 40^2)^{\frac{1}{2}}} = 8 \text{ A and lags } V_{BR} \text{ by } \phi = 53 \cdot 2°$$
$$\text{i.e., } I_3 = 8\angle -240° - 53 \cdot 2° = 8\angle -293 \cdot 2°$$

Using the above results the string phasor diagram is drawn.

The line currents are obtained graphically by applying the equations:

$$I_R = I_1 - I_3$$
$$I_Y = I_2 - I_1$$
$$I_B = I_3 - I_2$$

These are the consequence of Kirchhoff's First Law applied to points R, Y and B. Finally, the polar diagram is obtained by the usual method, although it is not strictly necessary for the solution of the problem. It does however show phase angles more clearly than the string diagram. Numerical value of $I_R = I_Y = I_B = 8\sqrt{3} = 13 \cdot 9$ A and each lags its respective line voltage by $53 \cdot 2° + 30° = 83 \cdot 2°$.

The above example is normally solved more quickly as follows:

Since the load is balanced then the current in each branch is

$$I = \frac{400}{(30^2 + 40^2)^{\frac{1}{2}}} = 8 \text{ A}$$

From the theory of the delta connection it is known that:

$$I_L = \sqrt{3} \, I_{phase} \text{ therefore } I_L = \sqrt{3} \times 8 = 13 \cdot 9 \text{ A.}$$

This is a current in each line conductor.

Phase angle of each phase current is

$$\phi = \tan^{-1} \left(\frac{40}{30} \right) = 53 \cdot 2°.$$

From the theory again it is known that each line current lags behind the phase current by 30° and therefore in this case each line current lags its line voltage by 30° + 53·2° = 83·2°.

When the loads are unbalanced it is necessary to find each phase current in turn and draw a phasor diagram to scale, because each may have a different value. Thus a shorter method of calculations is inapplicable.

2.34   Phase sequence measurement

One of many circuits which may be used to determine phase sequence of the three-phase supply requires a high resistance voltmeter taking a negligible current, a resistor and a capacitor. Ideally the value of R should be equal to that of the capacitive reactance X_C.

The circuit is shown in Fig. 2.48(a) marked with voltage and current arrows.

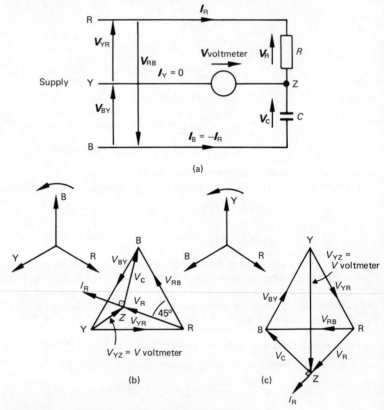

Fig. 2.48   Appertaining to phase sequence measurements

(i) *Standard sequence* R, Y, B. Direction: anticlockwise. The phasor diagram is shown in Fig. 2.48(b). The current I_R flowing through R and C is due to voltage V_{RB} and leads it by 45°. The voltages V_R and V_C are drawn in phase and lagging I_R respectively. The voltmeter V registers the voltage V_{YZ} which is small compared to the value of line voltage V_{RB}.

(ii) *Non-standard sequence* R, B, Y. The phasor diagram is drawn for this case in Fig. 2.48(c). The current I_R still leads V_{RB} but its position is now outside the triangle, and therefore the voltmeter reading V_{YZ} is high by comparison with the line voltage.

It must be noted that if the positions of resistor and capacitor are interchanged then the reverse results would be obtained. It is therefore necessary to draw the circuit carefully and also the phasor diagram, in order to be able to interpret the result correctly.

2.35 Power in three-phase a.c. circuits

The total average power is given by the sum of powers in each phase:

$$P = V_R I_R \cos \phi_R + V_Y I_Y \cos \phi_Y + V_B I_B \cos \phi_B \qquad (2.47)$$

If the load is balanced the equation (2.47) becomes:

$$P = 3 V_{ph} I_{ph} \cos \phi_{ph} \qquad (2.48)$$

For *star* connection $V_{ph} = \dfrac{V_L}{\sqrt{3}}$ and $I_{ph} = I_L$

substituting into (2.48):

$$P = 3 \frac{V_L}{\sqrt{3}} I_L \cos \phi_{ph} = \sqrt{3} \, V_L I_L \cos \phi_{ph}$$

For delta connection $V_{ph} = V_L$ and $I_{ph} = \dfrac{I_L}{\sqrt{3}}$

∴ substituting into (2.48)

$$P = 3 V_L \frac{I_L}{\sqrt{3}} \cos \phi_{ph} = \sqrt{3} \, V_L I_L \cos \phi_{ph}$$

Hence:

$$P = \sqrt{3} \, V_L I_L \cos \phi_{ph} \qquad (2.49)$$

Equation (2.49) applies to balanced loads irrespective of their configuration.

It must be emphasised however that the power factor is still a cosine of the angle between V_{ph} and I_{ph} phasors. The angle between V_L and I_L phasors is equal to $(30° \pm \phi_{ph})$.

SUMMARY

1. The *direct* current is the flow of electrons in one unchanging direction round the circuit.

The *alternating* current is the flow of electrons first in one direction and then in the reverse direction round the circuit.

D.C. circuits

2. Ohm's Law, $I = V/R$ $\qquad (2.1)$

3. Kirchhoff's current law,

$$I_1 \pm I_2 \pm I_3 + \cdots = 0 \qquad (2.2)$$

Kirchhoff's voltage law,

$$E_1 \pm E_2 + \cdots I_1 R_1 \pm I_2 R_2 + \cdots = 0 \tag{2.3}$$

4. Resistances in series

$$R_t = R_1 + R_2 + \cdots \tag{2.5}$$

5. Resistances in parallel

$$\frac{1}{R_e} = \frac{1}{R_1} + \frac{1}{R_2} + \cdots \tag{2.6}$$

6. Linear potential source is that which can be expressed by a straight line equation,
$V = E - Ir$ where V = terminal voltage in volts. $\tag{2.7}$

   E = e.m.f. of the source at no-load in volts
   r = internal resistance of the source in ohms
   I = load current in amperes.

7. Power in d.c. circuits $P = VI$ $\tag{1.4}$
   or $P = I^2 R$ $\tag{2.9}$
   or $P = V^2/R$ $\tag{2.10}$

Single phase a.c. circuits

8. Kirchhoff's laws apply without modification to alternating current circuits provided instantaneous values are used.

9. Sine wave p.d.'s and currents can be expressed symbolically by a phasor and their values written by a composite symbol:

$$\text{Magnitude} \angle \text{angle}$$

10. Kirchhoff's laws apply to phasor quantities but summation instead of algebraic must be geometric – usually done by drawing phasor diagrams to scale.

11. In a.c. circuits with sinusoidal excitation:

$$\text{The resistance: } R = \frac{V}{I} \tag{2.22}$$

and causes the p.d. and current waves to be in phase.

$$\text{The inductive reactance: } X_L = 2\pi f L = \frac{V}{I} \tag{2.25}$$

and causes the current to lag the p.d. by 90°.

$$\text{The capacitive reactance: } X_c = \frac{1}{2\pi f C} = \frac{V}{I} \tag{2.28}$$

and causes the current to lead the p.d. by 90°.

12. The combination of R, L and C in series gives the impedance

$$Z = \sqrt{(R^2 + (X_L - X_C)^2)^{\frac{1}{2}}} = \frac{V}{I} \tag{2.31}$$

and causes the current to lead or lag the p.d. by an angle ϕ less than 90°.

13. The combination of R, L and C in parallel gives the admittance Y

$$Y = \left(\left(\frac{1}{R} \right)^2 + \left(\frac{1}{X_c} - \frac{1}{X_L} \right)^2 \right)^{\frac{1}{2}} = \frac{I}{V} \tag{2.35}$$

and causes the total current to lead or lag the p.d. by an angle ϕ less than 90°.

14. Power in a.c. circuits $P = \text{(p.f.)} VI$ (2.38)
irrespective of the shape of the p.d. and current waveforms.

15. Power in a.c. circuits with sinusoidal excitation

$$P = VI \cos \phi \qquad (2.40)$$

where $\qquad \text{(p.f.)} = \cos \phi \qquad (2.41)$

Three-phase a.c. circuits

16. Three-phase supply is the result of the interconnection of three identical single phase a.c. sources, the p.d.'s of which are $120°$ out of phase with each other.

17. The three sources are usually connected in delta or in star or occasionally in open delta.

18. For delta connection,

Line voltage = phase voltage,

Line current = $\sqrt{3}$ phase current,

and *lags* phase current by $30°$.

19. For star connection,

Line voltage = $\sqrt{3}$ phase voltage

and *leads* phase voltage by $30°$.

Line current = phase current.

20. Two single phase supplies connected in open delta also produce a balanced three-phase supply.

21. Star connected supply produces either,

           (i) 4-wire system,

or   (ii) 3-wire system, when neutral wire is omitted.

22. Delta connected supply produces a balanced 3-wire system only.

23. Standard phase sequence is R–Y–B, i.e., anticlockwise. Non-standard phase sequence is R–B–Y or R–Y–B in a clockwise direction.

24. Power in three-phase unbalanced circuits equals the sum of powers in each phase.

25. Power in three-phase balanced circuits,

$$P = \sqrt{3} \ V_L I_L \cos \phi_{\text{ph}} \qquad (2.49)$$

EXERCISES

1. The diagram in Fig. 2.49 shows a simple d.c. potentiometer.

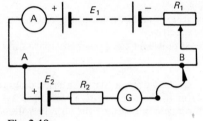

Fig. 2.49

$E_1 = 10$ V, $E_2 = 4$ V and $R_2 = 20 \ \Omega$

The internal resistances of the batteries are negligible. When the rheostat R_1 is adjusted with the sliding contact in position B to give zero deflection of the potentiometer the current read by the ammeter is 0·25 A. The sliding contact is then moved from B towards A through a distance equivalent to 2 Ω. Calculate the currents under these conditions giving their values and directions.

Answers: Current through E_1 = 0·242 A

Current through E_2 = 0·018 A

Current through wire AB = 0·26 A.

2. (*a*) Give at least three reasons why electrical energy is usually generated and transmitted in the form of alternating current rather than a direct current.

(*b*) Three e.m.f. sources A, B and C (Fig. 2.50) produce potential differences across their terminals given by equations:

$$v_A = \sqrt{2}\ 100 \sin 314t\ \text{(V)}$$
$$v_B = \sqrt{2}\ 150 \sin (314t + 60°)\ \text{(V)}$$
$$v_C = \sqrt{2}\ 180 \cos 314t\ \text{(V)}$$

Fig. 2.50

The reference directions of their potential differences are given by the arrows placed next to them.

Find graphically or by calculations their resultant voltage when they are connected:

(i) by joining terminals 2 to 3 and 4 to 5

(ii) by joining terminals 2 to 3 and 4 to 6

The resultant voltage should be expressed in the form

$$\sqrt{2}\ V_R \sin (314t \pm \phi).$$

Answers: (i) $\sqrt{2} \times 358 \sin (314t + 60°)$ (V)

(ii) $\sqrt{2} \times 186 \sin (314t - 16°)$ (V)

3. (*a*) Give at least two reasons why a sine wave was chosen as a basic wave for a.c. systems throughout the world.

(*b*) Show by means of a construction how a rotating line generates a sine wave.

How is such a line modified to obtain a phasor which adequately represents a.c. quantities such as voltage or current?

(*c*) A circuit connected to a supply, the voltage of which is $v = 353 \sin 314t$ V, takes a current $i = 7·07 \sin (314t + \pi/3)$ A.

Draw phasor diagram showing voltage and current in the circuit, and state whether it is predominantly inductive or capacitive.

Calculate the frequency of the supply and the power dissipated in the circuit.

Answers: 50 Hz, 625 W.

4. (*a*) For the circuit shown in Fig. 2.51 draw a 'string' phasor diagram assuming that the inductive reactance X_L is greater than the capacitive reactance X_C. From it draw a 'polar' phasor diagram. The reference condition of the supply voltage is given by an arrow in Fig. 2.51.

(*b*) If V = 400 V at 50 Hz, R_1 = 20 Ω, R_2 = 60 Ω, X_L = 70 Ω and X_C = 50 Ω

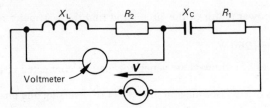

Fig. 2.51

Calculate
  (i) the current in the circuit
  (ii) the voltage read by the voltmeter shown in Fig. 2.51 assuming that voltmeter
       takes negligible current and
  (iii) the phase angle between supply voltage and the current.

Answers: (i) 4.85 A   (ii) 447 V   (iii) 14°2′.

5. For the circuit in Fig. 2.52, $R_1 = 30\ \Omega$, $R_2 = 20\ \Omega$, $\omega L = 40\ \Omega$, $1/\omega C = 40\ \Omega$
and supply voltage $V = 200 \angle 0°$ V.

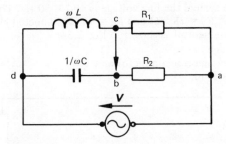

Fig. 2.52

Draw to scale a complete 'string' phasor diagram and find:

  (*a*) branch currents
  (*b*) current taken from the supply
and (*c*) the voltage between points 'c' and 'b', assuming the reference conditions
          shown by the arrow directed from 'c' and 'b'.

Answers: (*a*) $4 \angle -53°4′$ A, $4.47 \angle 63°26′$ A   (*b*) $4.5 \angle 10°$ A   (*c*) $180 \angle 101°$ V.

6. (*a*) In the circuit shown in Fig. 2.53 the supply voltage is given by the equation
$v = 300 \sin 314t$ V. Re-draw the circuit showing a wattmeter connected to it to
measure the power drawn from the supply and find the expression for the instantaneous
value of the current.

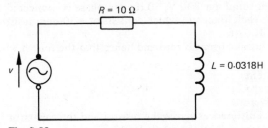

Fig. 2.53

(*b*) Sketch graphs of the voltage, current and instantaneous power marking clearly
portions of the latter which correspond to power transfer from the supply to the
circuit.

(c) Calculate the power factor of the circuit and find the wattmeter reading by two different methods.

Answers: 0·707 lag., 260 W.

7. (a) The windings of a three-phase generator may be connected either in star or in delta.

Show by means of a phasor diagram the effect of reversing one phase:

(i) if a star connection is used, and

(ii) if a delta connection is used.

State the relationship and displacement between the line and phase voltages so obtained.

(b) What value of current would circulate in the delta, case (ii), if the resistance per phase was 2 Ω, and the inductance per phase 0·011 H and the voltage per phase is 63·3 V at 50 Hz?

Answers: (b) 10·55 A.

8. (a) Obtain an expression for the power in a *balanced* three-phase circuit.

(b) In a three-phase, 4-wire system the line voltage is 400 V 50 Hz. The load consists of three resistors $R_1 = 23 \Omega$, $R_2 = 28·8 \Omega$ and $R_3 = 46 \Omega$ connected between the three line conductors and the neutral as shown in Fig. 2.54.

Find: (i) the current in each line,

(ii) the current in the neutral conductor and

(iii) the power in the load.

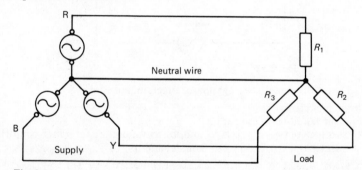

Fig. 2.54

Answers: (i) 10 A, 8 A, 5 A   (ii) 4·35 A   (iii) 5290 W.

9. (a) The windings of a 3-phase generator are connected in star. Show that for a correctly formed star the line voltage equals ($\sqrt{3}$ x phase voltage) and leads it by 30°.

(b) Such a generator supplying 230 V, 50 Hz per phase is connected to a balanced star-connected load, each branch of which consists of a 30-ohm resistor in series with a pure capacitor of 120 μF.

Show that the neutral current is zero and hence that the fourth wire can be dispensed with, and calculate

(i) the line current

(ii) the line voltage

(iii) the power factor

(iv) the potential differences across the resistor and the capacitor

(v) the power consumed by the whole load.

Draw the complete phasor diagram.

Answers: (i) 5·75 A   (ii) 400 V   (iii) 0·75 lead   (iv) 172·5 V, 152·2 V

(v) 2·98 kW.

10. (*a*) Define power factor and state for what circuits it is equal to the cosine of a phase angle between the voltage and the current.

(*b*) Three-phase 400-V, 50-Hz mains are connected to terminals, R, Y and B, to which the following apparatus is connected:

A resistance of 200 Ω between R and Y;

A choking coil of 0·636 H inductance and negligible resistance between Y and B; and

A capacitor of 15·9 μF capacitance between B and R.

Draw a phasor diagram showing:

   (i) the line voltages
   (ii) the currents in the circuits,
   (iii) the line currents,

and determine:

      (iv) the line currents
and (v)   the total power.

Answers: (iv) 3·86 A, 3·86 A, 2 A   (v) 800 W.

Principles of operation of machines

The operation of transformers and rotating electrical machines depends on a common set of fundamental principles. These principles apply to all types of transformers and to all a.c. and d.c rotating machines, irrespective of whether the latter operate as motors or as generators. In this chapter these principles are described in detail.

3.1   Fundamental principles

The operation of transformers and rotating electrical machines depends on the following phenomena:

1. The electric current always produces a magnetic field in the plane perpendicular to its direction of flow.
2. The magnetic field which is 'cut by' or 'cutting' the conductor causes an e.m.f. to be induced in it. (Law of induction.)
3. The interaction between two magnetic fields generates a force which tends to align them with one another. (Law of interaction.)

3.2   Magnetic field due to electric current

Figure 3.1 shows a conductor carrying an electric current which produces a magnetic field around it. The lines of magnetic flux are visualised as concentric circles in a perpendicular plane to the conductor. Their direction is given by that of the right-hand corkscrew rule as shown in the Fig. 3.1.

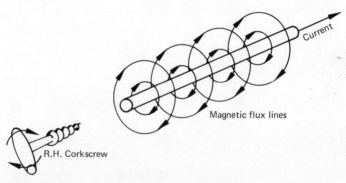

Fig. 3.1   Magnetic field due to electric current

3.3   The law of induction

Two forms of an e.m.f. of induction may be distinguished. The first is due to a movement of a conductor through the magnetic field. If the conductor moves in a straight
line the e.m.f. is designated by e_1; if the conductor rotates, it is designated by e_r and is
referred to as an 'e.m.f. of rotation'. The second, designated by e_p is due to pulsation
of the magnetic field. It occurs when the magnetic flux changes, that is, grows or
decreases in value through a stationary wire coil but where no mechanical movement
of parts takes place.

Both e.m.f.'s can occur at the same time in an electrical device, although in some
cases only one is present. A d.c. machine is an example of an e.m.f. of rotation only,
whereas a transformer is an example of an e.m.f. of pulsation only.

3.4   E.M.F. due to the motion of a conductor

When a conductor of length l meters is moved across a uniform magnetic field of
flux density B tesla with a constant velocity u meters per second, then the e.m.f.
e_1 volts induced in it is given by the equation:

$$e_1 = Blu \text{ (V)} \tag{3.1}$$

Figure 3.2(a) shows the arrangement in which the magnetic field is due to a current
of I amperes flowing in a wire wound around two magnetic poles. The conductor

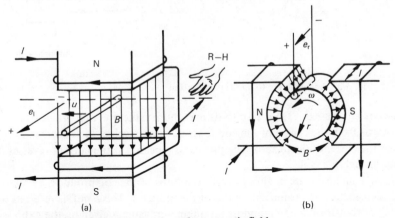

(a)   (b)

Fig. 3.2 Conductor moving through a magnetic field

'cuts' the flux at right angles to the magnetic lines and the direction of the e.m.f. is
found by Flemings Right-hand Rule, as indicated by the arrow in Fig. 3.2(a). The
Rule is applied as follows:

> The right hand is placed so that the magnetic lines emerging from the north pole
> enter the palm of the hand and the thumb points in the direction of motion. The
> outstretched fingers then indicate the direction of the induced e.m.f.

In electrical machines the linear motion is usually changed to a rotational motion so
that continuous generation of e.m.f. can be achieved. The conductor in Fig. 3.2(b) is
rotated at a constant velocity of ω rad/s in an anticlockwise direction, and its linear

velocity is $u = \omega r$ (m/s). Therefore instantaneous e.m.f. of rotation is given by:

$$e_r = Bl\omega r \text{ (V)} \tag{3.2}$$

The condition of cutting the flux at right angles is observed since the lines cross the air-gap by the shortest route, i.e. radially, and the air-gap is of uniform length under the pole shoes. The direction of the e.m.f. generated changes as the conductor rotates and in turn cuts the magnetic flux under the north pole and then the south pole.

3.5   E.M.F. of pulsation

An inductor fed with an alternating current i, which produces an alternating flux Φ, is shown in Fig. 3.3(a).

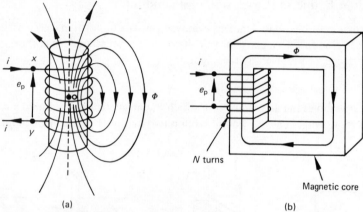

(a)   (b)

Fig. 3.3   Appertaining to e.m.f. of pulsation

As the current increases from zero to its maximum value, the flux grows from zero outwards from the centre of the coil to its maximum value, then decreases inwards as the current falls to zero. The process is repeated as the current reverses its direction through the coil.

The growth and decrease of flux through the turns of the coil produces an alternating e.m.f. across the end terminals x and y of the inductor. This e.m.f. is referred to as a pulsating or transformer e.m.f. and its instantaneous value is given by the expression:

$$e_p = \frac{\delta(\Phi N)}{\delta t} \text{ (V)}, \tag{3.3}$$

where $\delta(\Phi N)$ is a change (increase or decrease) of 'flux linkages' taking place in a short time interval δt seconds. The 'flux linkage' (ΦN) is the product of magnetic flux in webers and the turns of the coil through which that magnetic flux passes. The inductor shown has a non-magnetic core and the flux is not confined to the inside of the coil throughout its lengths but 'leaks out' more and more from the centre 'O' towards the ends x and y of the inductor. It is therefore necessary to multiply each portion of flux by the turns through which it passes and sum them up to arrive at the total value of (ΦN).

In the reactor shown in Fig. 3.2(b) a magnetic core is used in place of a non-magnetic one. It is much easier for the magnetic field to be 'established' in the core than in the air, therefore practically all the flux produced by a current is confined to the central core and links or goes through all the turns of the coil. Thus the number of turns which the changing flux goes through is always constant and only the change in its value has to be taken into account. 'N' therefore is taken outside δ sign and the equation (3.3) becomes:

$$e_p = N\frac{\delta\Phi}{\partial t} \text{(V)} \tag{3.4}$$

3.6 The law of interaction

The arrangement in Fig. 3.4(a) shows a conductor of length l meters placed in a uniform magnetic field of B tesla and carrying a current of i amperes.

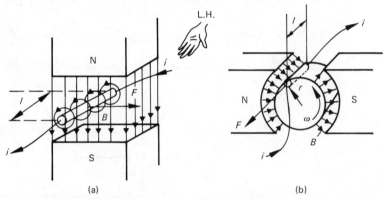

(a)   (b)

Fig. 3.4 Current-carrying conductor in a magnetic field

The current produces its own magnetic field which reacts with the main flux and as a result the conductor experiences a force in the direction perpendicular to that of the main magnetic flux lines. The value of the force F newtons, is given by the equation:

$$F = Bli \text{ (N)} \tag{3.5}$$

Whether the force acts to the left or to the right is found by Flemings Left-hand Rule. The Rule is applied as follows:

The left hand is placed so that the magnetic flux lines emerging from the north pole enter the palm of the hand and the outstretched fingers point in the direction of the current flow. The thumb then indicates the direction of the resulting force.

As with an e.m.f. generated, the simple linear case is adapted to produce continuous motion as shown in Fig. 3.4(b). The conductor is placed on a cylinder and supplied with a current i. The force experienced by it at a given instant is $F = Bli$ (N) as before, but its direction is at right angles to the radius r of the cylinder. The turning moment or torque on the conductor therefore is given by:

$$T = Fr. = Blir \text{ (Nm)} \tag{3.6}$$

It must be noted that to produce continuous motion, the direction of the current must be changed every time the conductor moves from the influence of one pole to the influence of the other.

3.7   Introduction to energy conversion

A simple electro-mechanical energy convertor is shown in Fig. 3.5(a). The switch 'S' is closed in position 1 and the coil moved from left to right across the magnetic field

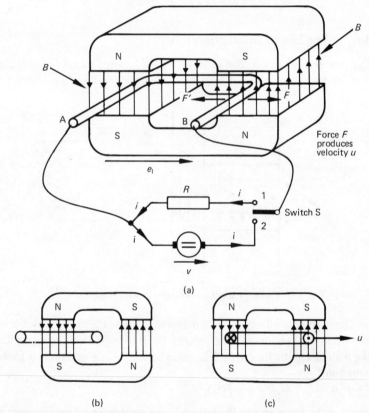

Fig. 3.5   Appertaining to electro-mechanical energy conversion

under the action of an external force. The following phenomena occur during the time interval it takes the coil to travel from the position shown in Fig. 3.5(b) to the position in Fig. 3.5(c).

1. As the conductors forming the two sides of the coil enter the magnetic field, the e.m.f. is induced in them. The e.m.f. drives the current round the coil and supplies the electrical energy to resistor R. Hence the electrical output of the coil has increased from zero by an amount δW_e (J).
2. When the coil moves through the magnetic field carrying a current, the two sides of the coil experience a force which opposes the motion. Therefore an externally applied force must increase to maintain the velocity at a value of

u (m/s). Hence the mechanical energy input is increased by an amount δW_m (J).

3. The coil accelerating from rest to velocity u (m/s) acquires a kinetic energy, i.e., it stores the mechanical energy, the amount of which is designated by δW_s (J).

4. The current flowing in the coil produces its own magnetic field and thus modifies the magnetic flux distribution in the air-gap between the two pairs of poles. Hence the energy stored in the magnetic field is changed by δW_f (J)

5. Various losses occur in the convertor such as i^2R loss in the coil, friction against air particles etc. The energy loss is denoted by δW_l (J).

The principle of *energy conservation* demands that the balance must be preserved between any two states of the system. Thus the energy equation can be written for the convertor considered between positions in Fig. 3.5(b) and (c) as follows:

$$\delta W_m = \delta W_e + \delta W_s + \delta W_f + \delta W_l \tag{3.7}$$

If the time taken by the coil in moving between the two positions is δt (s), then dividing equation (3.7) by δt, the *power* balance equation is obtained i.e.,

$$p_m = p_e + \frac{\delta W_s}{\delta t} + \frac{\delta W_f}{\delta t} + \frac{\delta W_l}{\delta t} \tag{3.8}$$

In this book the operation of rotating machines is considered under *steady state* conditions only, that is, at an instant when the mechanical or electrical load is constant.

Since the *transient* conditions of starting, stopping, or speed changing of the machine are not dealt with quantitatively, therefore, the change in stored mechanical energy of the system is zero,

i.e., $$\frac{\delta W_s}{\delta t} = 0$$

Similarly the conditions of changing the effective values of the currents which produce the magnetic field are not considered, and therefore the change in magnetic energy stored in the system is also zero,

i.e., $$\frac{\delta W_f}{\delta t} = 0$$

The power loss is not zero of course, and is taken into account in subsequent chapters.

In the convertor considered here however, the loss is neglected in order to clarify the mechanism of energy conversion,

i.e., $$\frac{\delta W_l}{\delta t} \text{ is neglected.}$$

Hence the equation (3.8) reduces to:

$$p_m = p_e \tag{3.9}$$

This expression is concerned solely with the region where the electro-mechanical energy conversion takes place.

3.8   Electromechanical power conversion – linear motion

GENERATOR PRINCIPLE

Let a force F(N) applied to a coil in Fig. 3.5(a) produce a motion at constant velocity u(m/s). Then the mechanical power p_m supplied to the coil is equal to the product of the force F and the velocity u, i.e.,

$$p_m = Fu \text{ (Nm/s)} \tag{3.10}$$

As the coil moves, both its conductors 'cut' the magnetic flux and an e.m.f. is generated between the points A and B.

The e.m.f. $e_1 = 2Blu$ (V). When the switch 'S' is in position 1, the coil is joined through flexible connections to a resistor R (Ω). Assuming the resistance of the coil, the switch and the connections to be negligible, the e.m.f. produces a current round the circuit given by

$$i = \frac{e_1}{R} \text{(A)}$$

Since the velocity of the coil is constant, the principle of 'equal action and reaction' demands that the force F causing the motion must be counter-balanced by an equal force, which is designated by F' in Fig. 3.5(a).

F' is due to a flow of the current i as can be verified by applying Fleming's Left-hand Rule to each conductor of the coil. From equation (3.5)

$$F' = 2Bli \text{ (N)}$$

and since $F' = F$

therefore $F = 2Bli \text{ (N)}$

Substituting in equation 3.10

$$p_m = 2Bliu = (2Blu)i = e_1 i = p_e$$

where p_e is an electrical power generated in the coil and supplied to the resistor R.

i.e., $p_m = p_e$

or

$$Fu \text{ (Nm/s)} = e_1 i \text{ (W)} \tag{3.11}$$

Hence in the ideal case considered, with no losses in the coil, the mechanical power supplied is equal to the electrical power generated as shown in the previous paragraph. This is the generator principle.

It must be noted that as long as the velocity of the coil remains constant, no change in the amount of energy stored in the magnetic field takes place. The field acts as a medium *through* which energy is transformed from *mechanical* to *electrical* form.

MOTOR PRINCIPLE

When the switch 'S' is in position 2, a source of v volts supplies the current of i amperes to the coil A – B.

Hence the electrical power p_e supplied to the coil is equal to the product of p.d. v and the current i,

i.e., $$p_e = vi \text{ (W)}. \tag{3.12}$$

Due to the current flow, each conductor experiences a force in the same direction, so that the total force on the coil is

$$F = 2Bli \text{ (N)}.$$

The coil moves from left to right as seen by application of Fleming's Left-hand Rule.

The motion involves cutting of the magnetic flux which in turn generates an e.m.f. $e_1 = 2Blu$, where u is the velocity of the coil and is assumed to be *constant*.

From Kirchhoff's voltage law, the e.m.f. is equal and opposite to the applied p.d. i.e.,

$$e_1 = v$$

therefore $$v = 2Blu \text{ (V)}$$

Substituting this in equation (3.12)

$$p_e = 2Blui = (2Bli)u = Fu = p_m$$

where p_m is the mechanical power generated by the coil in motion.

Again $$p_e = p_m$$

or

$$e_1i \text{ (W)} = Fu \text{ (Nm/s)}$$

The equation is identical with (3.11) except that the conversion is from *electrical* to *mechanical* power, through the medium of the magnetic field. This is the motor principle.

3.9   Electromechanical power conversion – rotational motion

The arrangement is shown in Fig. 3.6

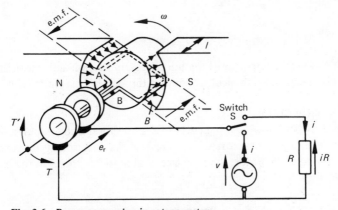

Fig. 3.6   Power conversion in rotary system

GENERATOR PRINCIPLE

A single turn coil which is placed on the cylindrical rotor is rotated by an externally applied torque T (Nm) at a constant velocity ω (rad/s) in a radial magnetic field of constant flux density B (T).

Hence the mechanical power p_m supplied to the coil is equal to the product of the torque T and the velocity ω, i.e.,

$$p_m = T\omega \text{ (Nm/s)} \tag{3.13}$$

As the coil rotates, both its conductors cut the magnetic flux and the e.m.f. of rotation is generated between the points A and B.

The e.m.f. $e_r = 2Bl\omega r$ (V)

When the switch 'S' is in position 1 the e.m.f. is applied to an external resistor R (Ω) via sliprings, brushes and external connections. The slip ring and its brush is a simple device for joining rotating and stationary parts of an electrical circuit. Assuming that the coil and all the connections have negligible resistance, then the value of the current is

$$i = \frac{e_r}{R} \text{(A)}$$

Since the velocity of rotation is constant the principle of 'equal action and reaction' demands that forward torque T must be counter-balanced by an equal torque which is designated by T' in Fig. 3.6.

T' is due to a force exerted on the two sides of the coil by the current flowing in the coil whilst it rotates through the magnetic field.

From equation (3.6) $T' = 2Blir$ (Nm)

and since $T' = T$

therefore $T = 2Blir$ (Nm)

Substituting in equation (3.13)

$$p_m = 2Blir\omega = (2Blr\omega)i = e_r i = p_e$$

where p_e is an electrical power generated in the coil and supplied to the resistor R.

i.e., $p_m = p_e$

or

$$T\omega \text{ (Nm/s)} = e_r i \text{ (W)} \tag{3.14}$$

Hence (3.14) is the important equation for rotary motion and shows the mechanical power being converted to electrical power, through the magnetic field.

MOTOR PRINCIPLE

If the resistor R is now replaced (switch 'S' in position 2) by a source of v volts, capable of supplying the current which reverses its direction at intervals equal to half the revolution of the rotor, then the electrical power supplied to the machine is:

$$p_e = vi = e_r i \text{ (W)}$$

since $v = e_r$ by Kirchhoff's voltage law.

If the resulting speed of the rotor is constant and equal to ω (rad/s) and the torque is T (Nm) then:

$$p_e = e_r i = 2Bl\omega ri = 2(Bli)r\omega = 2Fr\omega = T\omega = p_m$$

i.e.,
$$p_e = p_m$$

or
$$e_r i \text{ (W)} = T\omega \text{ (Nm/s)}$$

The electrical energy is converted to mechanical energy and the motor principle is seen in action. The equation is identical with (3.14) and the device just described is a reversible convertor. It transforms electrical energy to mechanical energy or vice versa.

In general therefore, under steady state conditions, the energy conversion region is concerned with e_r, i, T and ω, with the magnetic field acting as a conversion medium, with zero change in its stored energy. This is shown pictorially in Fig. 3.7.

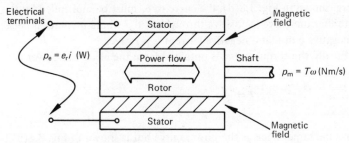

Fig. 3.7 Rotating machine – conversion region

3.10 Waveforms in a single coil machine

To consider all the phenomena occurring during one complete revolution the machine is redrawn in Fig. 3.8(a) in a *developed* form.

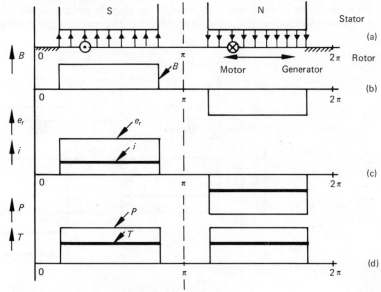

Fig. 3.8 Appertaining to machine in Fig. 3.6

In Fig. 3.8(b) the flux density distribution curve in the air-gap is plotted against the circumference of the rotor. The fringing of flux in the space between the poles is neglected, and since the air-gap is uniform the density is constant everywhere under the pole shoe. The flux, emanating from the north pole, is plotted as negative. Therefore that entering the south pole is positive. The waveform of the e.m.f. generated is shown in Fig. 3.8(c), and is seen to be an exact replica of the flux wave, since $e_r = 2Bl\omega r$ where l, ω, and r are constants. The e.m.f. e_r is therefore directly proportional to flux density B. As there are two conductors connected in series to form one turn, the maximum value of the e.m.f. is twice that generated in one conductor. The waveform of the current is also the same as that of the flux density curve because $i = e_r/R$. It is included in Fig. 3.8(c).

Since the instantaneous electrical power $p_e = e_r i$, therefore to obtain the power curve, shown in Fig. 3.8(d), the curve of e_r must be multiplied by the curve of i. Note that the curve is positive throughout the whole revolution because the product of negative e.m.f. and negative current is positive.

Finally the torque curve is obtained from the electrical power curve:

$$T = \frac{e_r i}{\omega}$$

It has the same shape as the power curve but is shown in Fig. 3.8(d) to a different scale.

QUESTION 1. A stator has two poles of arc equal to two-thirds of the pole pitch producing a uniform radial flux of density 1T. The length and diameter of the armature are both 0·2 m and the speed of rotation is 1500 rev/min. Neglecting fringing, draw to scale the waveform of e.m.f. generated in a single fully-pitched rotor coil of 10 turns. If the coil is connected through slip rings to a resistor of a value which makes the total resistance of the circuit 4 Ω, calculate the mean torque on the coil.

ANSWER.

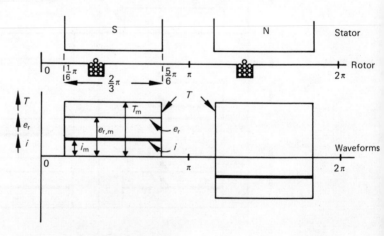

Fig. 3.9 Appertaining to Question 1

The maximum value of the e.m.f. of rotation:

$$e_r = Bl\omega r = 1 \times 0.2 \times \frac{2\pi \times 1500}{60} \times 0.1 \times 2 \times 10 = 20\pi = \underline{62.8 \text{ V}}$$

$$\therefore \qquad \text{Maximum } i = \frac{e_r}{R} = \frac{62.8}{4} = \underline{15.7 \text{ A}}$$

and max. torque
$$T = \frac{e_r i}{\omega} = \frac{62.8 \times 15.7}{\dfrac{2\pi \times 1500}{60}} = \underline{6.26 \text{ Nm}}$$

The average torque
$$= \frac{2 \times 6.26 \times 2\pi/3}{2\pi} = \underline{4.18 \text{ Nm}}$$

3.11   Single coil sinewave machine

Practically all electrical power is generated in the form of sinusoidal voltage and current and hence machines are built either to generate or to work from sinewave supplies.

The simple machine described in previous paragraphs and shown in Fig. 3.6 does not produce the sinewave e.m.f. It shows, however, clearly that the shape of the e.m.f. of rotation is directly proportional to the flux density curve, if the speed of rotation is kept constant.

It is therefore necessary to design a machine which gives sinusoidal distribution of magnetic flux density in the air-gap. One of a number of ways, is shown in Fig. 3.10(a), where the radial length of the air-gap is varied by shaping pole pieces so as to achieve this object.

The flux density waveform is shown in Fig. 3.10(b), and its equation is given by:

$$B_\theta = B_m \sin \theta \text{ (T)} \qquad (3.15)$$

If l is the length of each conductor

r the radius of the rotor carrying the coil, and

ω the rotational speed of the rotor, then

the e.m.f. induced in conductor A is obtained from equations (3.2) and (3.15), i.e.:

$$e_r \text{ (A conductor)} = B_m l\omega r \sin \theta = \sqrt{2}E \sin \theta \text{ (V)}.$$

Similarly e.m.f. induced in conductor B is

$$e_r \text{ (B conductor)} = B_m l\omega r \sin (\theta - \pi) = \sqrt{2}E \sin (\theta - \pi) \text{ (V)}.$$

The angle θ is decreased by π because the conductor B is placed $-\pi$ radians behind the conductor A.

In both cases $\sqrt{2}E$ replaces $B_m l\omega r$, where E (r.m.s.) per conductor $= \dfrac{B_m l\omega r}{\sqrt{2}}$ (V)

Expressing e.m.f. in each conductor in phasor form:

$$E_A \text{ (conductor 'A')} = E\angle 0°$$
$$E_B \text{ (conductor 'B')} = E\angle -\pi$$

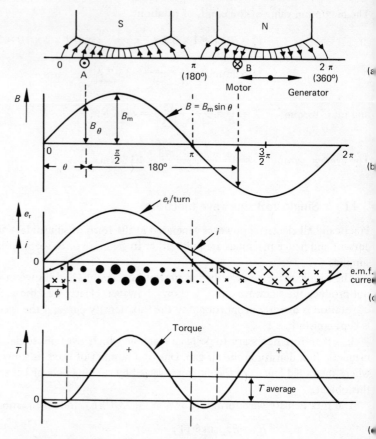

Fig. 3.10 Appertaining to sinewave machine

These are drawn in Fig. 3.11

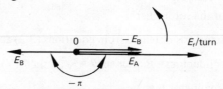

Fig. 3.11 Phasors of conductor e.m.f.'s

The e.m.f. in conductor A acts in reverse direction to that in conductor B, therefore the e.m.f. per turn is obtained from equation:

$$E_r/\text{turn} = E_A - E_B \tag{3.16}$$

Reversing E_B and adding it to E_A (Fig. 3.11)

$$E_r/\text{turn} = 2E \text{ per conductor.}$$

Therefore

$$E_r/\text{turn} = \frac{2B_m l \omega r}{\sqrt{2}} \text{ (V).} \tag{3.17}$$

The above equation gives an r.m.s. value of the e.m.f. generated in one turn of a coil for sinusoidal distribution of the magnetic flux. The e.m.f. waveform is a sinewave and is shown in Fig. 3.10(c).

The relative size of dots and crosses indicates the instantaneous magnitudes of the e.m.f. induced in the coil according to its position. For instance when the conductors A and B are passing exactly through the central line of each main pole, the e.m.f. in the coil is a maximum.

If the coil is now connected to an external load the current will be supplied by it. Its waveform will be a sinewave but its phase angle will vary with the nature of the load. For instance when a pure resistance is in the circuit the current wave will be in phase with the voltage wave. When the load is a combination of resistance and inductance then the current lags the e.m.f. by an angle ϕ. This condition is shown in Fig. 3.10(c). It is noteworthy that in this case the two conductors forming the coil carry a maximum instantaneous current when they *do not* cut the flux at maximum density. This point is indicated by dots and crosses displaced from those representing the e.m.f. e_r.

Finally the torque curve is plotted in Fig. 3.10(d). The curve is the product of e_r and i curves divided by a constant speed ω.

Therefore the average torque $= \dfrac{\text{average power}}{\omega}$

i.e., average torque $= \dfrac{E_r/\text{turn } I \cos \phi}{\omega}$

where I is the r.m.s. value of the load current and $\cos \phi$ is the power factor. The result may be interpreted by considering the machine generating electrical power for a portion of the cycle and motoring for the next portion of the cycle and so on. The mechanical primemover therefore drives the machine first, then is driven by it. Its net input is thus smaller than it would be if it had to drive it continuously.

3.12   Expression for e.m.f. of rotation in a single-coil machine

The equation (3.17) gives $E_r/$turn in terms of flux density B_m and the speed of rotation ω. It is however usual to express E_r in terms of its frequency and the total flux emanating from each pole, as follows:

The average flux density for half a cycle of a sinewave $= \dfrac{2}{\pi} B_m$ (T).

The area in which the flux is established under one pole $= \pi r l$ (m²)

total flux per pole $\Phi = \dfrac{2}{\pi} B_m \pi r l = 2 B_m r l$ (Wb)

and $B_m = \dfrac{\Phi}{2rl}$ (T)

Also if f is the number of cycles through which the e.m.f. changes in one second, then:

$\omega = 2\pi f$ (rad/s)

(1 revolution $= 2\pi$ radians)

Substituting these values in equation (3.17)

$$E_r/\text{turn} = \frac{2}{\sqrt{2}} \frac{\Phi}{2rl} 12\pi fr = \frac{2\pi\Phi f}{\sqrt{2}} \text{ (V)},$$

i.e.,

$$E_r/\text{turn} = 4{\cdot}44 \ f \text{ (V)} \tag{3.18}$$

The coil considered above is formed from two conductors spaced one pole pitch apart, that is from the central line of one pole to the central line of the next. Such a coil span is equal to 180 electrical degrees.

In most machines the coils are short-pitched, that is their span is less than $180°$.

This is done in order to improve the sinewave of the generated e.m.f. Sometimes it may also be dictated by the nature of the winding employed, of which a concentric type is an example.

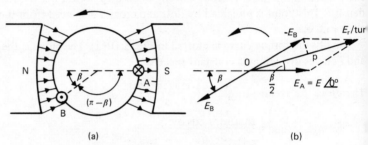

(a)   (b)

Fig. 3.12   Short-pitched coil and its phasor diagram

Fig. 3.12(a) shows the coil, the span of which is reduced by an angle of β electrical degrees. Thus the e.m.f. induced in the conductor B lags that in conductor A by an angle $(\pi - \beta)$. The e.m.f. generated in the complete turn is given by the equation (3.16) i.e.

$$E_r/\text{turn} = E_A - E_B,$$

which is solved graphically in Fig. 3.12(b).

From it the numerical value of E_r is:

E_r/turn of short pitched coil

$$= 2 \times 0 \text{ P} = 2E \cos\frac{\beta}{2}$$

$$= (E_r/\text{turn of fully pitched coil}) \times \cos\frac{\beta}{2}$$

Hence:

$$\cos\frac{\beta}{2} = \frac{E_r/\text{turn of short pitched coil}}{E_r/\text{turn of fully pitched coil}} = k_p \tag{3.19}$$

k_p is called the pitch factor and is always equal to or less than 1.

The equation (3.18) is finally modified by including k_p in it:

$$E_r/\text{turn} = 4{\cdot}44 \, k_p \Phi f \text{ (V)}. \tag{3.20}$$

It must be stressed that (3.20) applies only to either sinusoidally distributed flux in the air-gap or to its sinewave component.

QUESTION 2. Find the r.m.s. value of the generated e.m.f. in a single-turn coil short-pitched by 30 electrical degrees. The coil is rotated at 3000 rev/min in a two-pole stator which establishes sinusoidally distributed flux in the air-gap. Each pole produces the flux of 40 mWb.

ANSWER. Angle $\beta = 30°$

$\therefore$ pitch factor $k_p = \cos \dfrac{\beta}{2} = \cos \dfrac{30°}{2} = \cos 15° = 0.9659$

frequency $= \dfrac{3000}{60} = 50$ Hz.

The e.m.f./turn $= 4.44 \times 0.9659 \times 0.04 \times 50 = \underline{8.58 \ (V)}$

SUMMARY

1. The operation of transformers and rotating machines depends on:
   - (i) electric current producing a magnetic flux
   - (ii) the law of induction
   - (iii) the law of interaction.

2. Two e.m.f.'s of induction are distinguished

   (i) e.m.f. due to motion of a conductor, which is given as:
   $$e_1 = Blu \ \text{(V) for linear case} \tag{3.1}$$
   $$e_r = Bl\omega r \ \text{(V) for rotational case} \tag{3.2}$$
   where
   B = flux density in T
   l = conductor length in m
   u = linear velocity in m/s
   ω = rotational velocity in rad/s.
   and
   r = radius of the coil in m.

   (ii) e.m.f. of pulsation e_p which is given as:
   $$e_p = \frac{(\Phi N)}{\delta t} \ \text{(V) for non-magnetic core} \tag{3.3}$$
   $$e_p = \frac{N\delta\Phi}{\delta t} \ \text{(V) for magnetic core} \tag{3.4}$$
   where
   (ΦN) = flux linkage in weber-turns
   N = number of turns in the coil
   and
   t = time in seconds

3. Law of interaction:
   $$F = Bli \ \text{(N) for linear case} \tag{3.5}$$
   $$T = Fr = Blir \ \text{(Nm) for rotational case} \tag{3.6}$$
   where
   i = the current in A
   F = force in N
   and
   T = torque in Nm

4. Energy and power balance equations:
   $$\delta W_m = \delta W_e + \delta W_s + \delta W_f + \delta W_l \tag{3.7}$$
   $$p_m = p_e + \frac{\delta W_s}{\delta t} + \frac{\delta W_f}{\delta t} + \frac{\delta W_l}{\delta t} \tag{3.8}$$
   where
   δW_m = mechanical energy in J
   δW_e = electrical energy in J
   δW_s = stored mechanical energy in J
   δW_f = stored energy in magnetic field in J
   δW_l = loss energy in J.

5. Electromechanical power conversion

$$p_e = p_m \tag{3.9}$$

or $Fu = e_r i$ for linear case   (3.11)

and $T\omega = e_r i$ for rotational case   (3.14)

6. The waveform of e_r is the same as that of flux distribution curve in the air-gap, when the speed of rotation is constant. For sinewave e.m.f., the machines are designed to give as near as possible sinusoidal flux distribution in the air-gap i.e.,

$$B = B_m \sin\theta \text{ (T)} \tag{3.15}$$

7. E_r generated in one turn coil rotated in sinusoidally distributed flux is

$$E_r/\text{turn} = \frac{2}{\sqrt{2}} B_m l\omega r \text{ (V)} \tag{3.17}$$

or $E_r/\text{turn} = 4{\cdot}44\, K_p \Phi f$ (V)   (3.20)

where B_m = maximum flux density in T

Φ = total flux per pole in Wb

f = frequency of e.m.f. in Hz

and k_p = pitch factor.

8. Pitch factor $k_p = \cos\dfrac{\beta}{2}$ (3.19)

where β is an angle in electrical degrees by which the coils are short-pitched.

EXERCISES

1. Two coils A and B are arranged as in Fig. 3.13 and magnetically linked when either is energised. Coil A is wound on a hollow former and remains stationary. Coil B is wound on an iron core and can be moved into the hollow cylinder of Coil A. Describe what occurs in coil A when

   (i) Coil B is supplied from a battery and both coils remain stationary.

  (ii) Coil B is supplied from the battery and moved in and out of the coil A.

 (iii) Coil B is supplied from an a.c. mains at 50 Hz and both coils remain stationary.

Coil 'A'   Coil 'B'

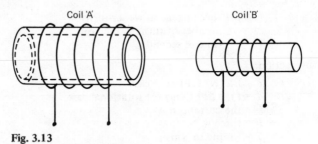

Fig. 3.13

Explain energy changes which take place and identify in each case,

  (a) the transformer principle, and

  (b) the electro-mechanical energy conversion.

2. A stator has two poles of arc equal to $\frac{5}{6}$ of the pole pitch producing a uniform radial flux. A cylindrical rotor has a single turn, fully pitched coil which is rotated in an anti-clockwise direction with the speed of ω rad/s.

Sketch the above arrangement and use it to deduce the expression

$$T\omega = e_r i$$

where T – is an instantaneous torque on the coil.

e_r – is an instantaneous e.m.f. of rotation generated in the coil.

i – is an instantaneous current flowing in the coil when the coil is connected through slip rings to an external circuit.

Give a short explanation of the electro-mechanical energy conversion using the expression deduced.

Also sketch the waveforms of e_r, T and i, if the load on the coil is purely resistive Neglect fringing.

3. A stator has two poles of arc equal to $\frac{3}{4}$ of the pole pitch producing a uniform radial flux of density 1·5 T. The length and diameter of the armature are both 0·2 m and speed is 1000 rev/min. Neglecting fringing draw to scale the waveform of the e.m.f. generated in a single fully pitched coil of 10 turns. If the average torque is 5 Nm find the average electrical power generated and the maximum current and sketch its waveform.

Answer: 523·5 W, 11·1 A.

4. A developed cross-section of a 2-pole machine with interpoles is as shown in Fig. 3.14. Sketch the waveform of e.m.f. generated in a single turn fully pitched coil and calculate the r.m.s. value of this e.m.f. given that: Flux density under the poles = 0·8 T; Flux density under interpoles = 0·16 T; Length of the coil side = 25 cm; Speed of rotation = 2400 rev/min.; Radius of the rotor = 5 cm. Neglect fringing.

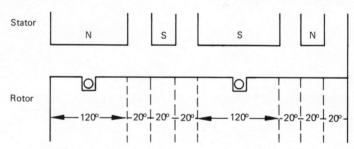

Fig. 3.14

Answer: 3·07 V.

5. A simple machine has a two pole stator and a cylindrical rotor. The faces of the stator poles are so shaped that the flux density in the air-gap is sinusoidally distributed, with a maximum of 1·2 T occurring at the centre of each.

Two slots one pole pitch apart are cut in the rotor, the length and diameter of which is 0·3 m. The rotor is driven at 3000 rev/min. A coil of 40 turns wound in the slots is brought out to two slip rings and connected to a resistor which brings the total resistance to 80 Ω.

(a) Sketch the arrangement and draw the waveforms of,
   (i) the e.m.f. and current generated in the coil,
   (ii) the torque exerted on the coil, stating the nature of the variation in each case.

(b) Calculate, (i) the r.m.s. value of the coil e.m.f. and current;
   (ii) their frequency, and
   (iii) the average torque.

Answer: 960 V; 12 A; 50 Hz; 36·4 Nm.

6. A 50-Hz generator has a flux of 0·1 Wb/pole sinusoidally distributed in its air-gap. Calculate the r.m.s. value of the e.m.f. generated in one turn which spans:

   (i) complete pole pitch;

   (ii) $\frac{5}{6}$ of a pole pitch.

Answer: 22·2 V, 21·4 V.

7. A four pole, single phase machine has 36 slots cut on its rotor. The machine is driven at 1500 rev/min and each of its poles produces a flux of 0·06 Wb.

   Calculate an e.m.f. generated in a single turn coil the sides of which are placed in two slots spaced

   (i) at 9 slots apart,

and (ii) at 8 slots apart.

Answer: 13·3 V; 12·9 V.

Main types of machines

In almost every country of the world the national power network either extends over the whole territory or is rapidly being developed.

In the British Isles the network is called the Grid and is shown in Fig. 4.1 together with the description detailing its various elements and indicating their functions.

The system is typical of most countries although different voltages may be used for its various sections.

This and succeeding chapters describe only the transformers and rotating machines, shown in Fig. 4.1. This group of devices may be characterised by the fact that their function is to convert either voltages and currents from one set of values to another (transformers) or electrical energy to mechanical energy and vice-versa (rotating machines).

4.1 Purpose and classification of transformers

The function of a transformer is to step up or down the voltage or current between two electrical circuits.

The equality of power on both sides requires that as the voltage is *stepped up* 'n' times, the current is stepped down 'n' times. This is not exactly so because a certain amount of power is lost in the transformer itself.

From the point of view of operation the transformers are classified as either

- (a) fixed voltage devices – the ratio of voltages remains constant for different loading conditions, or
- (b) fixed current devices – the ratio of currents is fixed for different loads on the transformer.

This classification is emphasised in Chapter 5 to underline the difference in operation. Furthermore the theory is confined to transformers of low frequency range (40-100 Hz) which have iron magnetic cores and does not deal with high frequency transformers.

The power frequency group can subsequently be divided according to application into:

- (i) power transformers
- (ii) distribution transformers
- (iii) testing transformers

and (iv) instrument transformers

Those in the first three groups work on the constant voltage principle.

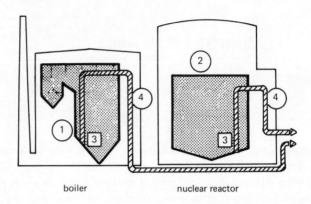

boiler   nuclear reactor

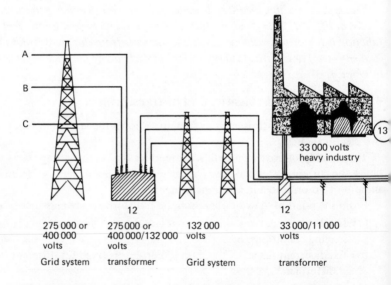

33 000 volts
heavy industry

275 000 or 400 000 volts	275 000 or 400 000/132 000 volts	132 000 volts	33 000/11 000 volts
Grid system	transformer	Grid system	transformer

Coal or oil is burned in the boiler **1** of a power station, or carbon dioxide gas is heated in
reactor **2** of a nuclear power station and the heat boils water circulating at high pressure ir
the boiler tubes **3** to create high-pressure steam **4**. The steam is taken by pipes to the turb
5 where it is used to drive the shaft **6** at high speed. From the turbine, the steam enters t
condenser **7** and passes over tubes containing cooling water. It is thus condensed back int
water and creates a vacuum which helps improve the flow of steam through the turbine. Th
water is returned to the boiler under pressure by a series of pumps.
The generator consists of a rotor **8** and a stator **9**. The rotor (an electro-magnet made of
number of windings mounted on a shaft) is coupled to the turbine shaft so that it is turned

Fig. 4.1   Electrical power system (*reproduced by courtesy of the Central
               Electricity Generating Board*)

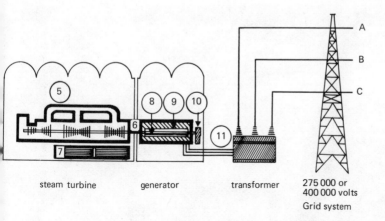

steam turbine   generator   transformer   275 000 or
                                                               400 000 volts

                                                               Grid system

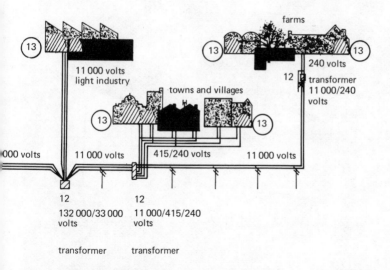

speed and generates electricity in more windings that make up the stator. A small genera-
10 driven from the end of the rotor shaft, produces the current required to energise the
r.

e largest modern generators electricity may be generated at about 25 000 volts but for
ient transmission over long distances the voltage is increased by transformers **11** to
000, 275 000 or 400 000 volts. The voltage is reduced again by other transformers **12**
istribution to consumers at suitable voltages—33 000 volts for heavy industries, 11 000 volts
ight industries and 240 volts for homes and farms. Different types of motors **13**.

Fig. 4.2(a) and (b) show two typical transformers belonging to groups (i) and (ii).

Fig. 4.2   (a) A 1000 MVA 400/275 kV power transformer.
     (G.E.C. Transformers Ltd.)

Fig. 4.2   (b) A 500 kVA 11 000/440 V distribution transformer.
(*G.E.C. Transformers Ltd.*)

Figure 4.3 shows a selection of transformers which work as constant current devices. Two basic forms of construction are illustrated. The first is the ring type with a bar primary and the second is the wound primary type.

4.2   Purpose and main types of rotating machines

The purpose of most electrical machines is electro-mechanical energy conversion. The machine which converts mechanical energy (energy of motion) into electrical energy is called a generator. The machine which converts electrical energy into mechanical energy is called a motor.

There is a tremendous number and variety of electrical machines being manufactured and in operation around the world. The machines vary in size from a few watts to millions of watts, but the majority of them can be grouped under the following four headings:

     (i) direct current machines
    (ii) induction machines
   (iii) synchronous machines
and (iv) polyphase commutator machines.

The last group although listed separately consists essentially of induction machines with extra commutator windings employed in a special way. This group is not dealt with in this text.

Fig. 4.3   Current transformers. (*E.E. Co.*)

The remaining types which do not fall within the above categories may yet be regarded as their composites.

For instance synchronous-induction machines combine the basic operational features of the second and third groups. The universal motor is really a series excited d.c. machine which operates from both a.c. and d.c. supplies.

4.3   Direct current machine

A typical d.c. machine is illustrated in Fig. 4.4(a). The rotor and stator are shown separately in Figs. 4.4(b) and (c) respectively. When the rotor is driven by a mechanical prime mover and a small amount of direct current electrical energy is

supplied to the winding on the stator, a large amount of direct current electrical
energy is generated in the winding of the rotor and taken from its commutator. The
machine acts as a *d.c. generator.* If direct current electrical energy is supplied to both
stator and rotor windings, the machine operates as a *d.c. motor* and produces
mechanical energy at the rotorshaft.

4.4   Induction machine

A large modern induction machine is shown in Fig. 4.5(a). Figure 4.5(b) shows a
completed stator, whilst the two main types of rotors used in induction machines are
illustrated in Figs. 4.5(c) and (d). The first rotor is a 'squirrel cage' type whilst the
second is of 'wound' construction with the windings brought out to the slip-rings.

When alternating current electrical energy is supplied to the stator winding, a portion
of it appears by 'induction' in the rotor winding. The rotor circuit may be closed
either internally as in the 'squirrel cage' type, or externally as in the wound type by
shorting three sliprings. A rotor torque results and the machine operates as an
induction motor within a specific speed, dependent on the frequency of the supply
and the machine construction.

(a)

(b)

(c)

Fig. 4.4   Typical screen protected d.c. machine in the range of
40 to 800 h.p. (*E.E.–A.E.I. Machines Ltd.*)

Fig. 4.5   (a) An 1100 kW, 4-pole, 11 000 V, 3-ph, 50 Hz closed air-circuit
water cooled cage induction motor with integral top-mounted
air-to-air heat exchanger. (*E.E.–A.E.I. Machines Ltd.*)

Fig. 4.5   (b) Wound stator of an 11 000 kW, 4-pole induction motor.
(*E.E.–A.E.I. Machines Ltd.*)

Fig. 4.5   (c) Cage rotor of an 11 000 kW, 4-pole induction motor.
           (*E.E.-A.E.I. Machines Ltd.*)

Fig. 4.5   (d) Wound rotor of a 9000 kW, 6-pole induction motor.
           (*E.E.-A.E.I. Machines Ltd.*)

If the machine running as a motor is 'speeded up' by mechanical means to run at a
higher speed than 'normal', electrical energy is generated in the stator winding and fed
back into the a.c. mains. It must be noted that the machine will act as an induction
generator when connected to the mains but will not normally generate when driven
in isolation. This is because neither winding is supplied with electrical energy when
the machine is not connected to the mains, and there is no magnetic field inside its
stator.

4.5   Synchronous machine

A very large modern synchronous machine (generator) is shown in Fig. 4.6(a) with cylindrical rotor being placed inside the stator.

Figure 4.6(b) shows a closer view of the stator with water cooled windings. Finally Fig. 4.6(c) shows an alternative construction of the rotor with projecting or 'salient poles', usually employed in much smaller machines.

The above illustrates two typical forms of rotor construction, the stator being similar in either case, although water cooling is used only for the largest machines in the range of 300 MW and upwards.

When the rotor winding is supplied with a small amount of direct current electrical energy and driven by a mechanical prime mover, a large amount of alternating current electrical energy is generated in the stator winding. The machine acts as an *a.c. generator.*

If the electrical energy is supplied to both the stator (a.c.) and the rotor (d.c.) windings, the machine operates as a *synchronous motor*, and produces mechanical energy at the rotor shaft.

Fig. 4.6   (a) 500 MW turbo-generator (*E.E.–A.E.I. Turbine Generators Ltd.*)

Fig. 4.6   (b) Stator core and windings for 500 MW generator
             (*E.E.-A.E.I. T.G. Ltd.*)

Fig. 4.6   (c) 4-salient pole rotor for 4000 h.p. synchronous machine.
             (*E.E.-A.E.I. Machines Ltd.*)

4.6   Polyphase commutator machine

The theory of these machines is outside the scope of this book, nevertheless, one of
many versions of a.c. commutator motors is illustrated in Fig. 4.7(a) to show their
essential similarity with the machines in other groups.

In Fig. 4.7(b) a rotor of Schrage type a.c. commutator motor is shown separately
and it is seen that it carries two windings. One winding terminates on three sliprings
and the other on the commutator, as indicated in paragraph 4.3.

(a)

(b)

Fig. 4.7 (a) **The construction of an a.c. commutator motor.** (*Laurence, Scott & Electro Motors Ltd.*) (b) **Typical rotor of an a.c. commutator motor.** (*E.E.-A.E.I. Machines Ltd.*)

4.7 Common features of rotating machines

A closer study of the machines illustrated in previous paragraphs shows that their common features can be listed as follows:

   (i) Each machine is reversible; that is it can be either a motor or a generator (there are in practice differences in construction to improve the performance for one or the other mode of operation).

  (ii) Every machine consists of two main parts; one stationary i.e., STATOR, the other rotating, i.e., ROTOR.

 (iii) The stator is a hollow cylinder with inner surface either smooth or with projections forming salient poles.

 (iv) The rotor is a cylinder with its outer surface either smooth or with projections forming salient poles.

  (v) Conductors which form windings are either distributed along the inner stator and outer rotor surface or placed around the salient poles.

 (vi) The windings carry electric currents which produce a common magnetic flux in the airspace between the stator and the rotor. The airspace is called an AIR-GAP

(vii) Each machine has appropriate devices for leading current in and out of its two main parts such as fixed terminals, sliprings, commutators, brushes etc.

(viii) Each machine has its supporting structure, enclosure, bearings, cooling arrangements and so on.

4.8 Fundamental differences between machine types

It is clear from the foregoing that all the rotating machines are very similar in appearance and general construction.

The *factor* which determines the type of a given machine is the MAGNETIC FLUX PATTERN produced in its air-gap. The magnetic field pattern in turn is created by the electric currents flowing in the stator and rotor windings.

Hence each type of machine can be completely described in terms of what KIND OF WINDINGS the machine employs and what KIND OF CURRENTS are lead into or taken from the windings. These facts, therefore, are used in defining each machine in subsequent chapters.

All other features which go to make up the individual machines, whilst very important, are not fundamental to the understanding of their operation.

4.9   Generalised electric machine

Figure 4.8 shows a small generalised electrical machine built for study, experimentation and demonstration. Such a machine can be run as all conventional d.c. machines, induction machines, synchronous machines, a.c. commutator machines, and others. It requires d.c. and a.c. supplies, both single and three phase.

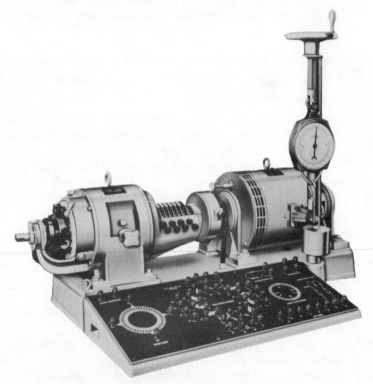

Fig. 4.8   Student demonstration set. (*Mawdsley Ltd.*)

All its stator coils and portions of rotor windings are brought out to the terminal board in order that different types of windings can be formed. This machine clearly demonstrates the essential unity underlying the theory of all rotating machines and enables the investigations to be carried out to show the differences between them.

SUMMARY

1. Typical electrical power system is shown in a diagrammatic form in Fig. 4.1. This textbook deals with transformers and rotating machines only.

2. The purpose of transformers is to step up or down the voltage or current between two electrical circuits.

3. The transformers operate either as constant voltage or constant current devices.

4. The purpose of a rotating electrical machine is an electromechanical energy conversion.

5. Most of the many electrical machines can be grouped as,

      (i) d.c. machines,
      (ii) induction machines,
      (iii) synchronous machines,
and   (iv) polyphase commutator machines.

By listing common features, it is seen that the rotating machines differ only according to the type of MAGNETIC FLUX PATTERN established in their air-gaps.

7. The flux pattern is produced by the currents flowing in different types of windings, hence the machines are defined in terms of,

      (i) types of windings used,
and   (ii) the kind of currents supplied to or taken from the windings.

8. Generalised machines illustrate both similarities and differences between different types of rotating electrical machines.

EXERCISES

1. Describe with the help of line diagrams the electrical supply system which is in operation in your country.

   State frequencies and voltage levels employed for

      (i) generation
      (ii) transmission
and   (iii) distribution of electrical energy.

2. (i) State clearly the purpose of an electrical transformer.
(ii) Indicate two different ways in which a transformer can be made to operate and give the names which are associated with each mode of working.

3. (i) What is the purpose of a rotating electrical machine?
(ii) Describe briefly the construction of a basic two pole machine which belongs to each of the following groups:

      (a) direct current machines
      (b) synchronous machines
and   (c) induction machines.

4. Enumerate common features to all rotating machines and state clearly the fundamental difference between them which determines the group to which each type belongs.

5. Write a short essay describing in general terms why and how it is possible to build a single generalised machine which can be made to operate as any one of the following types:

      (i) d.c. motor or generator
      (ii) synchronous motor or generator
      (iii) induction motor or generator.

Transformers

A transformer is a static device linking two electrical circuits, the function of which is to step-up or step-down an alternating voltage or current. A transformer may be operated as either constant-voltage or constant-current device, and so both modes of operation are considered here. This chapter describes the theory and operation of single phase and three-phase transformers, auto-transformers and moving coil voltage regulators.

5.1   The basic transformer

The simple transformer consists of two *electrical* circuits carrying alternating currents linked by a common *magnetic* circuit for the purpose of transferring electrical energy.

No energy conversion from electrical to other kind, such as mechanical or chemical, takes place in the transformer. Its function is to change the voltages and currents from the given values in the first circuit to the required values in the second circuit. Figure 5.1 shows the arrangement.

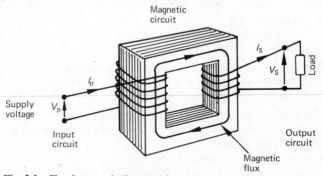

Fig. 5.1   Simple two-winding transformer

The input circuit consists of a coil called PRIMARY, wound around a magnetic core and connected to the alternating current supply. The output circuit consists of a similar coil termed a SECONDARY, wound around the *same* magnetic core and connected to the load.

Electrical power is supplied to the primary, transferred to the secondary via a magnetic flux which is established in the core and finally delivered to the load. The ratio of voltages and currents between the two circuits depends primarily on the number of turns in the two coils as will be explained in succeeding paragraphs.

5.2 Two modes of operation of a transformer

From the point of view of operation, transformers are classified as:

(i) constant voltage
and
(ii) constant current devices

Representing each winding by a coil symbol, Fig. 5.2(a) shows the first and, Fig. 5.2(b) the second type. In each case the transformer is loaded by a variable resistor.

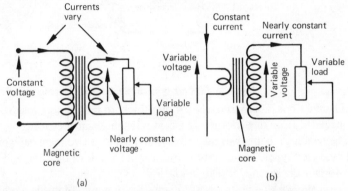

Fig. 5.2 (a) Voltage transformer (b) Current transformer. Two modes of
operation of a transformer

In the first type, the supply voltage is kept constant, giving nearly constant voltage in the secondary circuit for changing values of the current flowing through the load.

In the second type, the current through the first circuit is kept constant, giving nearly constant current through the load connected to the secondary circuit, thus varying voltage values across the load.

By far the greatest proportion of transformers in industrial use operate in the first manner, that is as 'voltage transformers'. The 'current transformers' are mainly used for measurements of large currents, operation of protective relays in supply systems and certain special applications.

The theory of both modes of operation is explained in detail in this chapter:

5.3 Practical forms of transformers

TWO-WINDING VOLTAGE TRANSFORMERS
Two methods of construction are usually employed. They are classified as (i) core type and (ii) shell type. Figure 5.3(a) shows the first arrangement and Fig. 5.3(b) the second.

In the core type the magnetic circuit consists of the two vertical legs or limbs with two horizontal sections, referred to as yokes. Each leg carries one half of both windings which are split in two equal sections.

In the shell type the magnetic circuit may be thought of as being 'wound' around the windings carried on its central portion. The windings are made up of flat coils arranged in a sandwich with primary and secondary interleaved.

In each of the two types the magnetic circuit is constructed from high grade silicon

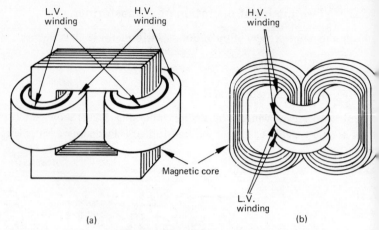

Fig. 5.3   (a) Core type (b) Shell type. Construction of voltage transformers

steel laminations, approximately 0·4 mm thick, to reduce core losses. The laminations are insulated from one another by a heat resistant enamel coating.

The windings themselves may also be divided into two main categories, i.e., tubular and sandwich types. The core type transformer employs the first, and the shell type transformer the second, as seen in Fig. 5.3.

The main working parts described are usually mounted inside a prefabricated steel tank and for larger transformers immersed in oil as a cooling medium. To increase the cooling surface area, the outside of the tank is fitted with corrugations or a large number of external pipes. The oil is air cooled either by convection or by fans. In very large units, air is replaced by water as a cooling medium.

TWO-WINDING CURRENT TRANSFORMERS
These transformers are also manufactured in two basic forms:

   (i) wound primary
   (ii) bar primary

Figure 5.4(a) shows the first and Fig. 5.4(b) the second.

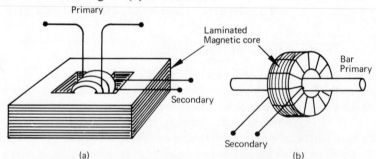

Fig. 5.4   (a) Wound primary (b) Bar primary. Construction of current
          transformers

The bar primary behaves as a single turn coil and is used for stepping large currents of hundreds of amperes down to a few amperes. The wound primary type is used for currents in the range of tens of amperes.

SINGLE WINDING AUTO-TRANSFORMERS

Figure 5.4 shows the circuit representing an auto-transformer. In this device one winding serves as both primary and secondary, but it is tapped at different points for input and output. Unlike the two-winding transformer, an auto-transformer transfers electrical power between two circuits partly through a magnetic link and partly by direct electrical connection.

The magnetic circuit may be of either core or shell construction.

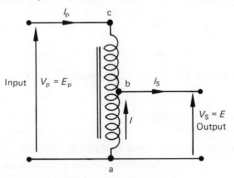

Fig. 5.5   Single phase auto-transformer

MOVING COIL TRANSFORMERS (REGULATORS)

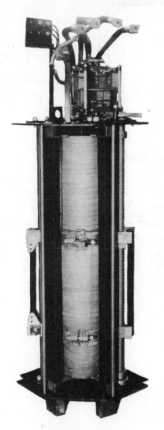

Fig. 5.6   Core and windings of single phase moving coil voltage regulator. (*Ferranti Ltd.*)

This is a special transformer with a movable short-circuited coil and is capable of varying output voltage smoothly over the predetermined range. Figure 5.6 shows the regulator.

The iron core is of multi-yoke construction, with four thin outside limbs surrounding the main central leg. The latter carries the principal winding, which consists of two halves wound in opposition to each other and joined in series. The third short-circuited coil of the same length as one half of the main winding is free to move up and down over it.

5.4   Operation of an ideal voltage transformer

Consider a transformer shown in Fig. 5.7. If the following simplifying assumptions are made:

   (i) the windings have negligible resistance
  (ii) the whole of the magnetic flux is confined to the magnetic core
 (iii) there are no hysteresis and eddy current losses in the core
  (iv) the magnetic flux is directly proportional to the ampere-turns producing it,

then the primary winding acting alone (switch S opened) behaves as a reactor with a pure inductance.

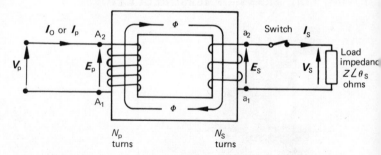

Fig. 5.7   Circuit of an ideal voltage transformer

The transformer terminals are marked according to British Standard 171, capital letters being used for the primary and lower case letters for the secondary. The two coils are so wound that at a particular instant of time both e.m.f.s act from terminals with lower to higher subscript numbers. Initially the switch 'S' is opened and an alternating p.d. of constant amplitude is applied to the primary. The p.d. drives a current which sets up an alternating flux in the core. The magnetic flux in turn links the turns N_p of the primary coil, induces in them an alternating e.m.f. e_p, which opposes v_p. Hence, by Kirchhoff's voltage law:

$$v_p = e_p$$

(resistance of the winding being negligible)

but,

$$e_p = N_p \frac{d\Phi}{dt} \text{ (see Chapter III equation (3.4)}$$

Thus assuming that the flux varies as a sinewave:

$$\Phi = \Phi_m \sin 2\pi ft \text{ (Wb)},$$

$$\therefore \qquad v_p = e_p = N_p \frac{d}{dt} \Phi_m \sin 2\pi ft = 2\pi f N_p \Phi_m \cos 2\pi ft,$$

or

$$v_p = e_p = 2\pi f N_p \Phi_m \cos 2\pi ft \text{ (V)}. \qquad (5.1)$$

The same magnetic flux also links the turns of the secondary coil and induces in them an e.m.f. e_s. The e.m.f. is shown acting from a_1 to a_2 because the current it would produce, if the switch S were closed, must produce an m.m.f., which would oppose the original flux. This is demanded by Lenz's Law of 'Action and Opposite Reaction'.

Hence $\qquad\qquad e_s = N_s \dfrac{d\Phi}{dt}$

and substituting for Φ,

$$e_s = 2\pi f N_s \Phi_m \cos 2\pi ft \text{ (V)} \qquad (5.2)$$

Examination of the two equations shows that when the flux Φ varies as a sinewave, both e.m.f.s, e_p and e_s are cosinewaves (sinewaves shifted forward by $90°$) and applied p.d. v_p must also be sinusoidal for it is equal to e_p. Putting $t = 0$ in equation (5.1) and (5.2) the maximum values of e_p and e_s are:

$$E_{p,m} = 2\pi f N_p \Phi_m \cos 0° = 2\pi f N_p \Phi_m \text{ (V)}$$

and $\qquad\qquad E_{s,m} = 2\pi f N_s \Phi_m \cos 0° = 2\pi f N_s \quad_m \text{ (V)}$

Dividing by $\sqrt{2}$ gives their r.m.s. values, i.e.

$$V_p = E_p = \frac{2\pi}{\sqrt{2}} f N_p \Phi_m = 4{\cdot}44 f N_p \Phi_m \text{ (V)} \qquad (5.3)$$

and

$$E_s = \frac{2\pi}{\sqrt{2}} f N_s \Phi_m = 4{\cdot}44 f N_s \Phi_m \text{ (V)} \qquad (5.4)$$

Figure 5.8(a) shows a phaser diagram for the no load condition (S opened), at an instant of time $t = 0$. V_p, E_p and E_s have maximum values and are drawn vertically, the flux Φ_m is zero at that instant and, therefore, its phasor is drawn horizontally lagging the e.m.f.s by $90°$.

The current I_o flowing through N_p turns produces an m.m.f. of $I_o N_p$ ampere-turns which is responsible for maintaining the magnetic flux. Although the current I_o is not a sinewave because in practice the B/H curve of the magnetic core is not a straight line, nevertheless it is possible to replace it by its equivalent sinewave and thus include it in the phasor diagram in Fig. 5.8(a).

The power supplied to the ideal transformer on no-load is given by $V_p I_o \cos 90°$ $= E_p I_o \cos 90° = 0$ in agreement with the assumptions (i) and (iii).

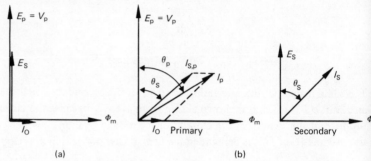

(a)   (b)

Fig. 5.8   (a) No-load condition (b) On-load condition. Ideal transformer phasor diagrams

When the equation (5.3) is divided by equation (5.4), then

$$\frac{E_p}{E_s} = \frac{N_p}{N_s} \tag{5.5}$$

Thus e.m.f.s are *exactly* in the ratio of turns of the two windings.

When the switch S is now closed; the e.m.f. E_s causes a current I_s to flow through the load. Its value is determined by the load impedance: $Z\angle\theta_s$, i.e.

$$I_s = \frac{E_s}{Z} \text{ amperes and its angle } \theta_s \text{ is assumed to be lagging.}$$

This current produces an m.m.f. $I_s N_s$ which opposes the m.m.f. $I_o N_p$ responsible for setting up the original flux Φ_m. From equation (5.3) it is seen that the flux Φ_m cannot decrease while the supply p.d. is kept constant and, therefore, the primary current increases from I_o to a new and larger value I_p, such that:

$$I_p N_p - I_s N_s = I_o N_p \tag{5.6}$$

Since the currents are not in phase the equation must be solved graphically and not by ordinary algebra. If the *increase* in primary current is now denoted by $I_{s,p}$ (secondary current I_s reflected in the primary winding), so that:

$$I_p - I_{s,p} = I_o, \tag{5.7}$$

then multiplying both sides by N_p,

$$I_p N_p - I_{s,p} N_p = I_o N_p. \tag{5.8}$$

Comparing (5.6) with (5.8) gives

$$I_{s,p} N_p = I_s N_s,$$

because the other two terms are identical in both equations. Solving for $I_{s,p}$,

$$I_{s,p} = \frac{N_s}{N_p} I_s \tag{5.9}$$

which when substituted in equation (5.7) gives:

$$I_p - \frac{N_s I_s}{N_p} = I_o \tag{5.10}$$

The graphical solution of equation (5.10) is shown in phasor diagram of Fig. 5.8(b). Diagrams for primary and secondary circuits are drawn separately for clarity, with E_p and E_s being parallel. These two e.m.f.'s are, in step (in phase) with each other because both are produced by the *same* common flux Φ_m

I_p is obtained by graphical addition of I_o and $I_{s,p}$ (equation (5.7)). The above explanation is based on the principle of 'constant flux' which is the consequence of applied p.d. to the primary remaining constant throughout the variation in the impedance of the load: Usually the current I_o is very small compared with the full value of I_p, ($I_o = 0.1 I_p$ to $0.05 I_p$).

Hence, $I_{s,p}$ is approximately equal to I_p and the equation (5.9) can be rewritten as:

$$I_p \simeq \frac{N_s I_s}{N_p} \tag{5.11}$$

From the above it is clear that the currents in the primary and secondary circuits are nearly inversely proportional to the ratio of their own turns.

QUESTION 1. An ideal single phase transformer is connected to the 240-V, 50-Hz. supply. Its maximum flux in the core is 0·003 Wb. The transformer is required to produce a secondary p.d. of 50 V. Calculate the number of turns in the primary and secondary windings and state the voltage per turn for each coil.

ANSWER. For the ideal transformer $V_p = E_p$. Using equation (5.3).

$$N_p = \frac{V_p}{4 \cdot 44 \, f \Phi_m} = \frac{240}{4 \cdot 44 \times 50 \times 0 \cdot 003} = \underline{360 \text{ turns}}$$

voltage per turn $= \dfrac{240}{360} = 0 \cdot 67 \text{ V}.$

Using equation (5.5).

$$N_s = N_p \frac{E_s}{E_p} = 360 \times \frac{50}{240} = \underline{75 \text{ turns}}$$

Voltage per turn $= \dfrac{50}{75} = \underline{0 \cdot 67 \text{ V}}.$

The voltage per turn in both windings is the *same*, as would be expected from the fact that the common flux is responsible for its production.

QUESTION 2. A 10-kVA, 400/200-V, 50-Hz single phase transformer delivers full load current at 0·8 p.f. lagging. If its no load current is 3 A, find the primary current and its power factor. Assume that the core losses, and voltage drops due to resistances and reactances are negligible.

ANSWER. Low voltage full load current

$$I_s = \frac{10\ 000}{200} = 50 \text{ A}.$$

$$I_{s,p} = 50 \times \frac{200}{400} = 25 \text{ A}.$$

The phasor diagram is drawn to scale according to Fig. 5.8(b).

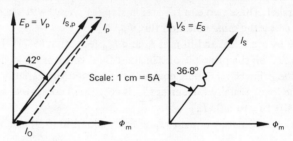

Fig. 5.9. Phasor diagrams for Question 2

From the diagram

$$I_p = \underline{27 \text{ A}}$$

and the power factor $= \cos 42° = \underline{0.742}$ lagging

5.5   Operation of a practical voltage transformer

Figure 5.10(a) shows the equivalent circuit of a practical transformer which includes winding resistances, leakage fluxes and the magnetic core, omitted in the treatment of the last paragraph.

In Fig. 5.10(b) the leakage flux is replaced by two pure inductive coils of reactance $X_p = 2\pi f L_p$ and $X_s = 2\pi f L_s$ ohms. This is possible because:

    (i) each leakage flux links one winding only,

    (ii) it is caused by the current in that winding alone,

and   (iii) one half of its path lies in the air which has a constant and very large reluctance compared with the reluctance of the magnetic core.

In order to simplify the solution of transformer problems without any significant loss of accuracy, it is possible to transfer the resistance and leakage reactance of one winding to the other and combine them into single values for each quantity.

Figure 5.10(c) shows the transformer with transfer effected from primary to secondary winding. The new resistance and reactance are:

    R_T = the total resistance of both windings

    X_T = the total leakage reactance of both windings.

Considering no-load operation when the switch 'S' is opened, the equations are obtained by applying Kirchhoff's voltage law for the primary circuit:

$$V_p = E_p \qquad\qquad\qquad\qquad\qquad (5.12)$$

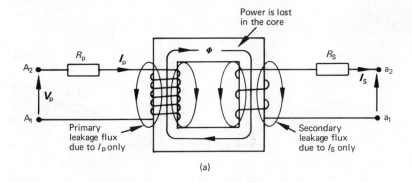

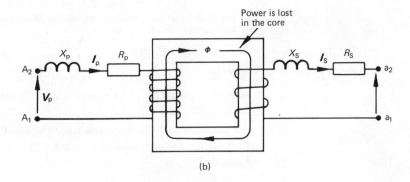

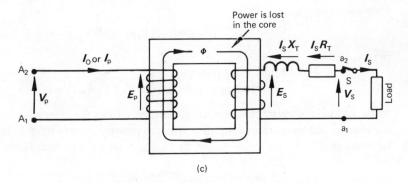

Fig. 5.10 Circuit of a practical transformer

and for the secondary circuit:

$$E_s = V_s \qquad (5.13)$$

The phasor diagram for this condition is shown in Fig. 5.11(a). It differs from that for an ideal transformer in that the no-load current I_o lags the supply p.d. V_p by an angle θ_o which is smaller than $90°$. This is because the core absorbs the power equal to $V_p I_o \cos \theta_o$. The power, which appears as heat is termed core loss and is due to hysteresis and eddy currents.

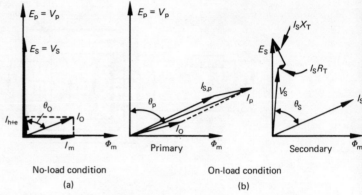

Fig. 5.11 Practical transformer phasor diagrams (a) No-load condition
(b) On-load condition

It is usual to resolve the no-load current I_o into two components I_m and I_{h+e} as shown in the phasor diagram, where

I_m = magnetising current, which is responsible for setting up the magnetic flux Φ_m, and

I_{h+e} = loss current which is responsible for supplying the power dissipated in the core as hysteresis and eddy current loss.

Thus I_m $= I_o \sin \theta_o$

and $I_{h+e} = I_o \cos \theta_o$

Therefore

$$\text{Core loss} = V_p I_o \cos \theta^\circ = V_p I_{h+e} \text{ (W)} \tag{5.14}$$

and

$$\begin{aligned}\text{Magnetising (reactive)}\\ \text{Volt-Amperes} = V_p I_o \sin \theta^\circ = V_p I_m \text{ (VA)} \end{aligned}\tag{5.15}$$

When the load is connected to the transformer by closing the switch S, it receives a current of I_s amperes. This in turn increases the primary current from I_o to a new value denoted by I_p. Applying Kirchhoff's voltage law to both circuits, the equations are:

For the primary circuit

$$V_p = E_p \tag{5.16}$$

and for the secondary circuit

$$E_s - I_s R_T - I_s X_T - V_s = 0 \tag{5.17}$$

These are solved graphically in Fig. 5.11(b). The phasor diagrams are drawn starting with the secondary circuit, where V_s, I_s and the phase angle between them θ_s, are known from the value of the impedance of the load. E_s is then obtained by adding to phasor V_s, $I_s R_T$ in phase and $I_s X_T$ 90° ahead of I_s. Magnetic flux Φ_m lags E_s by 90° and serves as a reference for both diagrams.

The primary phasor diagram has $E_p = V_p$ drawn in phase with E_s and I_o lagging it by an angle θ_o°. $I_{s,p}$ is scaled down according to equation (5.9) and drawn in parallel with I_s.

Finally I_p is the sum of $I_{s,p}$ and I_o. The phase angle θ_p between V_p and I_p is always slightly greater than θ_s making the primary power factor ($\cos \theta_p$) smaller than the power factor of the load. In most transformers the winding resistances and leakage reactances are kept very small and therefore if the supply p.d. V_p remains constant, V_s changes only slightly with increasing load current. The transformer operates as a substantially constant voltage device.

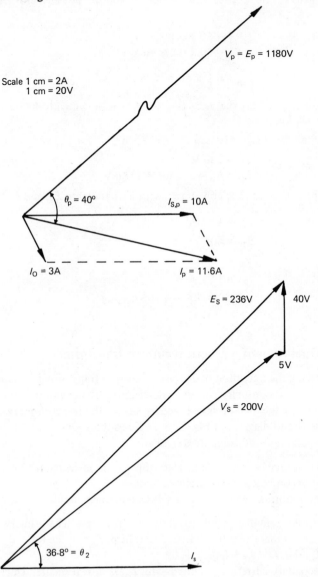

Fig. 5.12 Phasor diagrams for Question 3

QUESTION 3. The following data refers to a single phase 50-Hz transformer:

> no load current = 2 A
> no load p.f.   = 0·3 lagging
> total resistance transferred to secondary winding = 0·1 Ω
> total leakage reactance transferred to secondary winding = 0·8 Ω
>
> turns ratio $\dfrac{N_p}{N_s}$ = 5

Find the primary supply p.d., current and the power factor if the transformer is to supply a load at 200 V, with a current of 50 A at 0·8 power factor lagging.

ANSWER. $I_s R_T = 50 \times 0\cdot1 = 5$ V
$I_s X_T = 50 \times 0\cdot8 = 40$ V

$$I_{s,p} \; = \frac{1}{5} \times 50 = 10 \text{ A}$$

The phasor diagram for the secondary winding is drawn with the load current as a reference i.e. horizontal.

From secondary phasor diagram

E_s = 236 V

and $V_p = E_p = 5 \times 236 = \underline{1180 \text{ V}}$

Primary phasor diagram is now drawn with E_p parallel to E_s, and I_o lagging it by $\theta_o = \cos^{-1} 0\cdot3 = 72\cdot5°$

From it I_p = 11·6 A

and θ_p = 40°

hence $V_p = \underline{1180 \text{ V}}$

I_p = $\underline{11\cdot6 \text{ A}}$

and   p.f.= $\cos 40° = \underline{0\cdot766}$ lagging

5.6   Operation of a practical current transformer

In considering the operation of a constant current transformer it is important to remember that the primary current is determined by the external circuit and therefore the variation of the load on the secondary side of the transformer *does not* affect its value. The circuit diagram in Fig. 5.13(a) shows the arrangement.

Two operating conditions are described:

(i) the normal; when the secondary supplies an instrument of negligible impedance i.e. secondary is short circuited, and

(ii) the abnormal; when secondary is left on open circuit.

The phasor diagram in Fig. 5.13(b) shows the first case. The primary current I_p is constant and is therefore drawn first. It produces I_o and the reflected secondary current $I_{s,p}$, i.e. $I_p = I_o + I_{s,p}$.

The secondary current I_s, in phase with $I_{s,p}$ is shown smaller, since current transformers are invariably step down devices. $I_s R_s$ and $I_s X_s$ are the potential

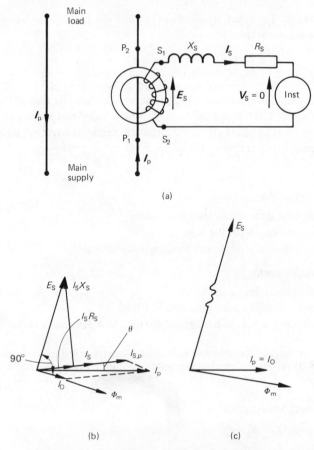

Fig. 5.13 (a) Circuit diagram (b) Phasor diagram with secondary on short circuit (c) Phasor diagram with secondary on open circuit. Current transformer circuit and phasor diagrams

differences across resistance and leakage reactance of the secondary winding and vectorially add up to e.m.f. E_s.

In practice, especially in bar primary transformers the resistance and leakage reactance of the primary circuit are so small that they are neglected altogether. Hence, R_s and X_s are, in fact, the same as R_T and X_T for current transformers. The secondary voltage V_s is almost equal to zero because the instrument requires negligible p.d. to drive the current through. Finally the magnetising flux Φ_m is drawn $90°$ lagging behind E_s. It must be emphasised that since E_s is small, Φ_m is small and so is I_o responsible for establishing the flux in the core. The difference between I_p and $I_{s,p}$ is solely due to the presence of I_o. Thus every effort is made to decrease it so that the ratio of the currents is as nearly equal to turns ratio as possible.

When the impedance of the load is increased, the voltage V_s must now increase so that I_s is still of the same value as before, because I_p remains unchanged. This necessitates higher E_s, and larger I_o. The extreme situation is reached when the secondary is opened. The e.m.f. E_s increases enormously in an effort to establish I_s required by the relationship

$$\frac{I_p}{I_s} \sim \frac{N_s}{N_p}$$

Since this is not possible, the whole of I_p produces the magnetic flux giving very high values of E_s. (Fig. 5.13(c)). This condition must be avoided and secondary of the current transformer must always be short circuited, unlike the voltage transformer where short circuit produces very high currents.

Summarising:

For voltage transformers:

    (i) The potential differences remain constant.
    (ii) The currents vary with the load.
    (iii) The short-circuit currents rise to dangerous values.

For current transformers:

    (i) The currents remains constant.
    (ii) The potential differences vary with the load.
    (iii) The open circuit p.d. of the secondary rises to a dangerous value.

QUESTION 4. A current transformer with a bar primary is required to step 800 A down to 5 A. If its no load current is negligible, calculate the number of turns required on the secondary.

ANSWER. Since I_o is negligible

then
$$\frac{I_p}{I_s} = \frac{N_s}{N_p}$$

and
$$N_s = N_p \times \frac{I_p}{I_s} = 1 \times \frac{800}{5} = \underline{160 \text{ turns}}$$

5.7   Operation of an auto-transformer

The circuit of an auto-transformer is shown in Fig. 5.5. By applying Kirchhoff's current law to point 'b'.

$$I_p - I_s + I = 0,$$

and solving for I.

$$I = I_s - I_p \tag{5.18}$$

If N_p is the number of turns between points 'a and c', N_s is the number of turns between points 'a and b', and neglecting no-load current, then the ampere-turns due to current I_p must be compensated by ampere-turns due to current I.

i.e.
$$I_p(N_p - N_s) = N_s I$$

substituting (5.18) for I

$$I_p(N_p - N_s) = N_s(I_s - I_p)$$

$$I_pN_p - I_pN_s = N_sI_s - N_sI_p$$

and

$$I_pN_p = N_sI_s$$

or

$$\frac{I_s}{I_p} = \frac{N_p}{N_s}$$

Similarly if the resistances and leakage reactances of the winding are neglected then

$$V_p = E_p \text{ and } V_s = E$$

hence from (5.5)

$$\frac{V_p}{V_s} = \frac{N_p}{N_s}$$

therefore

$$\frac{V_p}{V_s} = \frac{I_s}{I_p} = \frac{N_p}{N_s} \qquad (5.19)$$

The combined equation shows that the ratios of voltage and current depend on the number of turns just as in a two-winding transformer. The differences, however, between an auto-transformer and a two-winding transformer of the same rating are as follows:

   (i)   The no-load current is very much smaller in an auto-transformer
  (ii)   The leakage reactances and resistances are smaller in an auto-transformer
 (iii)   The amount of conducting material used is less because the common portion of the winding in an auto-transformer carries the difference between I_s and I_p
 (iv)   The whole of the auto-transformer winding must be insulated for the higher voltage
  (v)   There is a direct connection between the primary and the secondary sides.

An auto-transformer has greatest advantage over a two-winding transformer if the voltage ratio does not exceed 2:1, and there is no need to separate electrically high and low voltage sides. It is used extensively in laboratories as a variable voltage device, the secondary tapping being made in the form of a brush which can slide along a narrow strip of winding where the insulation has been removed. In single phase distribution lines it is employed to boost the voltage at consumer terminals.

5.8   Operation of a moving coil voltage regulator

The operation of a single phase regulator can be explained by reference to its diagrams of connections (Fig. 5.14) with movable coil 'S' in two limiting positions and marked with voltage and current arrows according to the convention.

The two fixed coils 'A' and 'B' are always connected in series opposition and joined to the supply. The third coil 'S', which is short circuited upon itself is free to move up or down the iron leg carrying fixed coils. The output voltage is taken from across coil 'A'. In Fig. 5.14(a) the coil 'S' surrounds the upper coil 'A', whereas in Fig. 5.14(b) it surrounds the lower coil 'B'.

Considering the first arrangement, the current I_p drawn from the supply causes each fixed coil to establish its own flux. The flux Φ_A due to the upper coil induces a current I in coil 'S', which in turn produces the flux Φ_s in opposition to Φ_A (Lenz's Law). Thus the resultant magnetic flux linking coil 'A' is very small and its impedance is mainly due to its ohmic resistance.

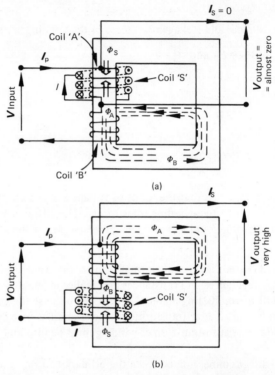

Fig. 5.14 Circuit diagram for M.C. voltage regulator

The coil 'B' on the other hand is surrounded by a large alternating flux Φ_B, which gives it a large self-inductance L_B and hence large reactance $X_B = 2\pi f L_B$ in addition to its small ohmic resistance.

The input voltage applied to both fixed coils produces potential differences across them according their relative impedances. Hence with the coil 'S' in top position the p.d. across coil 'A' is very small and that across coil 'B' very large. The output voltage is therefore very small.

When the coil 'S' is moved downwards until it surrounds coil 'B' (Fig. 5.14(b)) the impedance of coil 'A' is increased to its maximum value whereas that of coil 'B' is reduced to a very small value equal to its ohmic resistance. The output voltage is then very large and almost equal to the supply p.d.

Intermediate values of output voltage can be obtained by placing coil 'S' in between these two limiting positions. This arrangement enables the p.d. to be varied *smoothly* over the range of almost 0% to 100% of the supply voltage. Such regulators are used in laboratories in conjunction with testing equipment or in special applications. However the largest number of M.C. Voltage Regulators is found in power supply systems, where their purpose is to maintain constant voltage output when the input varies with different loading conditions.

The voltage of transmission and distribution lines usually changes by 10% to 25% of its normal rated value and therefore the regulators need only to provide the variation within these limits.

Such regulators often have an additional fixed coil 'R' placed above the coil 'A' and connected as shown in Fig. 5.15(a) and (b). The coil depending on the way it is wound (or connected) can either inject a buck (reducing) or boost (increasing) voltage in series with the line.

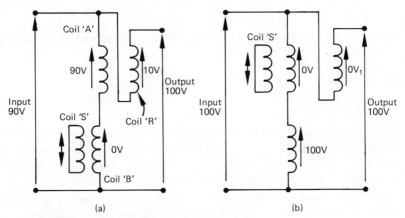

Fig. 5.15   M.C. voltage regulators for supply lines

In Fig. 5.15(a) and (b) the coil 'A' is assumed to have nine times as many turns as coil 'R', and the latter is wound to boost the line voltage on the output side. Simple calculations show that the output voltage will be maintained at 100 V when the input varies between the limits of 90 V and 100 V (the effect of winding resistance is neglected).

If the connection of coil 'R' is now reversed with respect to coil 'A', then its voltage would be subtracted from the input voltage and the regulator would produce 100 V output for input variation of 100 V to 112·5 V.

When both buck and boost are required a fourth fixed coil similar to coil 'R' is added at the bottom of the leg and below coil 'B'. It is connected in series with the other three fixed coils. In this arrangement 'R' acts as a boosting whereas the bottom coil as a bucking one for different positions of the moving coil 'S'.

It must be noted that although coil 'S' acts as a shortcircuited secondary of a two winding transformer, nevertheless its current I is very small because of its high impedance. Hence its presence does not increase materially the regulator's no-load current. Furthermore the waveform of the no-load current is practically sinusoidal and constant for all positions of the moving coil. This is due to low operating flux density and the air-gap in the magnetic circuit as seen in Fig. 5.14(a) and (b). The losses in the regulator are much the same as in an ordinary auto-transformer with additional resistive loss in the coil 'S'.

The position of a moving coil can be adjusted manually or by means of a small induction disc motor. The coil itself is clamped to a non-magnetic frame carrying a nut which moves up or down on a square threaded shaft. The shaft itself is rotated by hand or by a motor.

5.9   Losses in a transformer

In any transformer the power is lost in the resistance of the windings and in the magnetic core. The first is called an I^2R *loss* whilst the second a *core loss.*

I^2R loss is given by:

$$P_r = I_S^2 R_T \text{ (W)} \tag{5.20}$$

and varies with the load in a voltage transformer, but remains substantially constant in a current transformer.

The core loss is due to

    (i) eddy currents

and (ii) hysteresis.

Figure 5.16(a) shows a portion of a transformer leg made up of three laminations. The magnetic flux which links the turns of the coils passes through the core, the sections of which act as a series of complete turns. The e.m.f. is induced in each according to equation $E = 4\cdot44f\Phi.$ (V) which drives the circulating currents as shown. These are called 'eddy currents' and they produce power loss in the resistance of all the magnetic paths. To reduce this loss the core is made up of very thin laminations so that resistance is increased and therefore the value of eddy currents reduced.

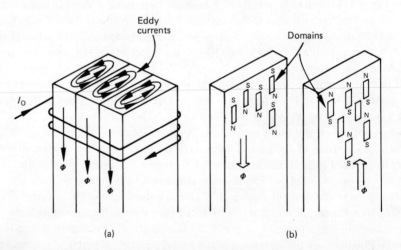

Fig. 5.16   (a) Eddy currents (b) Hysteresis. Magnetic core losses

The hysteresis loss is due to alternate magnetisation of the atoms, forming domains, in the magnetic material of the core. Each domain behaves as a very small magnet which aligns itself with the magnetic flux in which it is placed. As the flux changes its

direction the small magnets rotate to and fro as shown in Fig. 5.16(b). The power is
expended in this process and appears as heat.

If $\qquad$ P_c = total core loss

$\qquad\qquad\qquad$ P_h = hysteresis loss

and $\qquad$ P_e = eddy currents loss

then

$$P_c = P_h + P_e = V_p I_o \cos \theta_o \text{ (W)} \qquad\qquad (5.21)$$

Since the voltage transformers operate from constant voltage supply the core loss is
constant because the magnetic flux does not change by more than 2% between no
load and full load. In a current transformer this loss varies with changes of the load
connected to the secondary which in turn requires increase in magnetic flux.

5.10   Estimation of losses from practical tests

The 'voltage' transformer losses may be obtained by performing:

(i) an open circuit test—which gives the power wasted in the magnetic core, and
(ii) a short circuit test—which gives the power lost in the resistance of both wind-
$\qquad$ ings, and enables the values of R_T and X_T to be estimated.

5.11   Open circuit test

The transformer is connected to normal voltage supply with secondary open circuited.
Figure 5.17 shows the diagram of connections and includes all the necessary instruments.

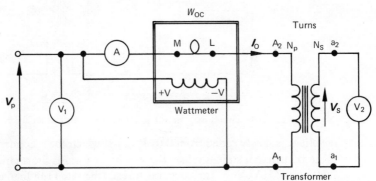

Fig. 5.17   Circuit diagram for O.C. test on a transformer

The voltmeter V_1 reads the normal operating voltage V_p (V). The ammeter A reads
the no-load current I_o (A). The voltmeter V_2 reads secondary voltage V_s (V), and the
wattmeter reads the power consumed by the transformer W_{oc} (W).

The power lost in the transformer is due to core losses and a very small resistive
loss in the resistance of the primary winding through which I_o flows. Since the no
load primary current is approximately $\frac{1}{20}$th of the full load value of I_p the resistive
loss is $\frac{1}{400}$th of its full load value and can be neglected. The core loss however depends

on the magnitude of flux which in turn is determined by the supply voltage V_p according to equation (5.3).

Since the test is performed at operating voltage, the core loss is nearly the same at no load as at full load, and therefore the wattmeter reads the core loss P_c only, i.e. $W_{oc} = P_c$. The no-load power factor can also be calculated from the expression (5.21).

$$\text{no-load p.f.} = \cos\theta_o = \frac{P_c}{V_p I_o} \qquad (5.22)$$

as well as the turns ratio:

$$\frac{N_p}{N_s} = \frac{V_p}{V_s}$$

5.12   Short circuit test

In this test one winding, usually 'Low Voltage' one, is shorted through an ammeter and low resistance connections. The other winding (H.V.) is joined to a variable voltage supply. The circuit diagram is shown in Fig. 5.18 with all the instruments necessary for the test.

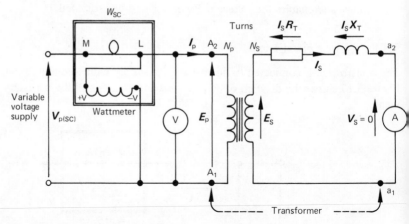

Fig. 5.18   Circuit diagram for S.C. test on a transformer

The voltage is slowly raised from 0 to $V_{p(sc)}$ at which the secondary current I_s attains the transformer's rated value. The voltmeter V reads short circuit supply voltage $V_{p(sc)}$ (V) on the H.V. side. The ammeter A reads the rated full load current on the L.V. winding and wattmeter the power W_{sc} (W) which is dissipated by the transformer alone, since it does not supply an external load.

This power loss is made up of $I^2 R$ loss in the winding resistances and a small core loss. The latter is proportional to Φ_m which in turn is determined by the applied p.d. The $V_{p(sc)}$ is very small in the test since the e.m.f. E_s need only overcome the potential differences across R_T and X_T, without providing an output voltage ($V_s = 0$). Thus the wattmeter registers resistive loss only, or

$$W_{sc} = P_r = I_s^2 R_T \text{ (W)}$$

from which the total winding resistance R_T referred to the L.V. winding is calculated, i.e.

$$R_T = \frac{P_r}{I_s^2} \ (\Omega.) \tag{5.23}$$

Furthermore from the readings of the S.C. test the value of X_T can also be calculated as follows:

since

$$Z_T = \frac{E_s}{I_s}$$

where

$$E_s = \frac{N_s}{N_p} E_p = \frac{N_s}{N_p} V_p \ (sc)$$

and

$$Z_T = \sqrt{R_T^2 + X_T^2}$$

therefore

$$X_T = \sqrt{Z_T^2 - R_T^2} \ (\Omega) \tag{5.24}$$

In the above equation all values are known or can be obtained from this test except the turns ratio N_p/N_s, which is calculated from the O.C. test results.

QUESTION 5. A 50-kVA, 400/230-V, 50-Hz, single phase transformer gave the following test results:

No-load test $V_p = 400$ V $I_o = 8$ A $P_c = 640$ W

Short-circuit test V_p (sc) $= 38$ V $I_s = 218$ A $P_r = 712$ W

Find the values of

    (i) turn ratio
    (ii) no-load p.f.
    (iii) total resistance and leakage reactance of the windings transferred to the L.V. side.

ANSWER.

    (i) turn ratio $\dfrac{N_p}{N_s} = \dfrac{400}{230} = \underline{1 \cdot 74}$

    (ii) no load p.f. $\dfrac{640}{400 \times 8} = \underline{0 \cdot 2}$ lagging

    (iii) $Z_T = \sqrt{R_T^2 + X_T^2} = \dfrac{E_s}{I_s} = \dfrac{38 \times (1/1 \cdot 74)}{218} = \underline{0 \cdot 1} \ \Omega$

            $R_T = \dfrac{712}{(218)^2} = \underline{0 \cdot 015} \ \Omega$

and $X_T = \sqrt{0 \cdot 1^2 - 0 \cdot 015^2} = \underline{0 \cdot 099} \ \Omega$

5.13   Efficiency of a transformer

The general expression for efficiency (1.7) is given in Chapter 1. Substituting in it in terms of symbols of previous paragraphs the per unit efficiency of the transformer at any value of the load is obtained.

$$\text{efficiency } \eta = 1 - \frac{I_s^2 R_T + P_c}{V_s I_s \cos \theta_s + I_s^2 R_T + P_c} \text{ p.u.} \qquad (5.25)$$

where $V_s I_s \cos \theta_s$ = transformer output in watts.

The maximum value of the efficiency occurs when the $I^2 R$ losses are equal to core losses, i.e. when

$$I_s^2 R_T = P_c = V_p I_o \cos \theta_o.$$

Expression (5.25) applies to:

     (i) two winding transformers
     (ii) auto-transformers
and   (iii) voltage regulators.

In the case of current transformers efficiency is usually of secondary importance as compared with their current ratio accuracy. Nevertheless the same expression can be used for its calculations if need be.

QUESTION 6. Find the efficiency of a 10-kVA, 1000/200-V single phase transformer supplying a full load at 0·8 p.f. lagging. Its total resistance transferred to secondary = 0·1 Ω and its no-load current measured on the H.V. side is 2 A at 0·3 p.f. lagging.

ANSWER.   Output = 10 000 x 0·8 = 8000 W

$$\text{Full load secondary current} = \frac{10\ 000}{200} = 50 \text{ A}$$

Resistive loss = 50^2 x 0·1 = 250 W

Core loss = 1000 x 2 x 0·3 = 600 W

$$\text{efficiency} = 1 - \frac{850}{8850} = 1 - 0\cdot096 = \underline{0\cdot904} \text{ p.u.}$$

5.14   Characteristics and regulation of voltage transformers

It has been stated that when a transformer operates from a constant voltage supply, it delivers *nearly* constant voltage output to the load. The question that is of importance to the user is of course 'how nearly' constant is the voltage under different loading conditions.

The answer is obtained experimentally or by calculation from the phasor diagram shown in Fig. 5.11(b). In the former method the load connected to the transformer's secondary is varied both in magnitude and power factor and the terminal voltage V_s and the load current I_s measured. The graphs of V_s are then plotted against I_s at different power factors giving the *load characteristics* for the transformer under test. Curves in Fig. 5.19 are typical for a large power transformer.

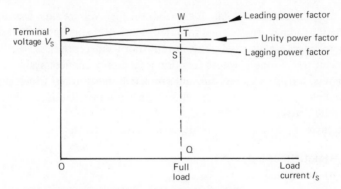

Fig. 5.19 Voltage transformer load characteristics

It is seen that the voltage V_s drops slightly with increasing values of I_s for unity and lagging power factors, whereas it actually increases above its no-load value when the power factor is leading.

The numerical variation between no-load and full load voltage is called inherent regulation, i.e.

$$\text{inherent voltage regulation} = V_s \text{ (no-load)} - V_s \text{ (on load)} \quad (5.26)$$

Its value can be obtained from the load characteristic. For example the regulation at unity power factor = OP − QT. It is often convenient to express regulation as the ratio of voltage variation to its no-load value, thus

$$\text{regulation} = \frac{V_s \text{ (no load)} - V_s \text{ (on load)}}{V_s \text{ (no load)}} \text{ p.u.} \quad (5.27)$$

The per unit regulation is useful to application engineers, for the limits of voltage variations are specified in this form. For instance transformers in public supply system must be so adjusted that the voltage at consumers terminals must not exceed ± 0·06 of its nominal value.

How to obtain the regulation by means of a phasor diagram in Fig. 5.11(b) is illustrated by worked example No. 3 and the following question:

QUESTION 7. Calculate:

    (i) Inherent voltage regulation

and (ii) per unit voltage regulation for the transformer in question 3.

                terminal voltage V_s = 200 V at a load of 50 A, 0·8 p.f. lagging

                terminal voltage $V_s = E_s$ = 236 V at no load.

∴   inherent voltage regulation = 236 − 200 = 36 V

and   regulation = $\dfrac{36}{236}$ = + 0·1525 p.u.

5.15   Characteristics of the current transformers

The principal use of transformers operated as current devices is to step large alternating currents down to small values in order to supply measuring instruments or protective relays, and to isolate these from the high voltage side. Usually the secondary windings are designed for rated values of 5 amperes, although 1 and 2-ampere ranges are also used. The current transformers are classified either as instrument transformers, whose function is to supply ammeters and current coils of wattmeters, watt-hourmeters etc. and protective transformers which supply various relays employed in protection systems of electrical installations. Both groups are covered by British Standard 3938.

The current transformers introduce two kinds of errors:

      (i) ratio (current) error,

and (ii) phase angle error.

Both errors are due to no-load current I_o which is necessary to establish magnetic flux in the transformer's magnetic core. Whilst it is impossible to eliminate the errors, nevertheless it is important that they are kept within limits prescribed by the British Standard. The ratio error is usually given as 'percentage ratio error' (%R.E.) and is defined in B.S. 3938 as:

$$\% \text{ R.E. at } I_p = \frac{K_n I_s - I_p}{I_p} \times 100\% \qquad (5.28)$$

where I_p is actual primary current

     I_s is actual secondary current

and K_n is nominal current ratio.

i.e.,
$$K_n = \frac{I_{pn}}{I_{sn}} = \frac{\text{Nominal primary current}}{\text{Nominal secondary current}}$$

For instance if the C.T.'s nominal ratio K_n = 100/5 A i.e. 20, then when 99 amperes produce secondary current of 5 A,

$$\text{then the } \% \text{ R.E.} = \frac{20 \times 5 - 99}{99} \times 100\% = \frac{100}{99} = 1{\cdot}01\%.$$

The phase angle error is measured in degrees and minutes, and is the angle between I_p and I_s shown on the phasor diagram in Fig. 5.13(b). Thus the current transformer characteristics are the curves of both errors plotted to the base of percentage secondary current. If the secondary winding is rated for 5 amperes then 5 is taken as 100 per cent. The shapes of error curves illustrated in Fig. 5.20 are characteristic of most current transformers.

The 'ratio error' graph shows that at small values of the current I_s the error is negative, then it decreases to zero, becomes positive and finally increases steadily as I_s approaches 100%.

The phase angle error is high at low values of I_s and steadily decreases with secondary current increase. By convention the phase error is taken as positive when I_s leads I_p, and since this occurs under almost all operating conditions, the 'phase angle error curve' lies wholly above zero axis.

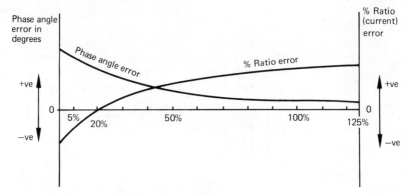

Fig. 5.20 Current transformer characteristics

5.16 Transformers for three-phase supplies

To step the voltage up or down in a 3-phase supply system two main methods are available; one employs three identical single phase transformers; and the other, one 3-phase transformer.

The three transformers are referred to as a 3-phase bank, and are shown in Fig. 5.21. In this arrangement each primary winding maybe connected across two lines of the 3-phase supply voltage. The secondaries are then connected either in star or delta to form three output terminals. Each transformer operates in exactly the way explained in previous paragraphs.

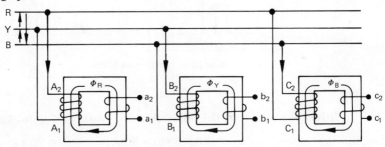

Fig. 5.21 Three-phase transformer bank

A 3-phase transformer unit on the other hand consists of a common magnetic circuit, but still retains separate windings for each phase. Construction of the magnetic core maybe understood by considering three single phase transformers, positioned at $120°$ to each other and shown in Fig. 5.22(a).

The limbs carrying primary windings are on the outside, whereas the inner limbs are placed close together at the centre. The resultant flux in the adjacent legs is the sum of the individual fluxes, which add up to zero as shown in the phasor diagram (Fig. 5.22(b)). Hence if the top and bottom yokes are joined together, the three central limbs are not necessary and can be dispensed with.

In order to simplify the construction and produce more compact design the three remaining limbs are arranged in line, with primary and secondary windings placed on each.

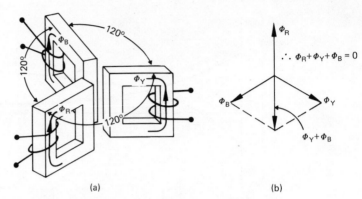

Fig. 5.22 Development of a three-phase transformer core

This arrangement shortens the yoke of the central leg. The unbalance of fluxes and consequently of magnetising currents introduced by it is nevertheless slight.

The resulting construction gives the 3-limb, 3-phase transformer, most commonly built in this country. For the largest units five limb construction is often adopted in order to reduce the overall height. Both types of cores are shown in Fig. 5.23(a) and (b) respectively.

It is clear that in a five limb core, the yoke carries only half the flux established in the main limbs, and therefore its cross-section is only half of theirs.

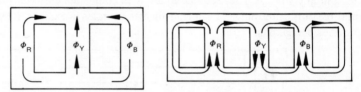

Fig. 5.23 Three-phase transformer cores

5.17   Interconnection of the three-phase transformer windings

The primaries and secondaries of either 3-phase bank or a 3-phase transformer may be connected in three different basic ways, namely:

> (i) Star,
> (ii) Delta,
> or   (iii) Zig-zag.

The first two are dealt with in Chapter 2. The third connection is obtained by dividing each secondary phase winding into two equal sections and interconnecting as shown in Fig. 5.24(a).

The phasor diagram in Fig. 5.24(b) is obtained as follows: The circuit is marked according to the convention with winding voltage arrows pointing towards terminals with higher subscript numbers, i.e. from a_1 to a_2, a_3 to a_4, etc.

Assuming that 3-phase voltages are induced in the windings with the phase sequence a, b, c, then voltages a_1-a_2 and a_3-a_4 are in phase with each other, but 120°

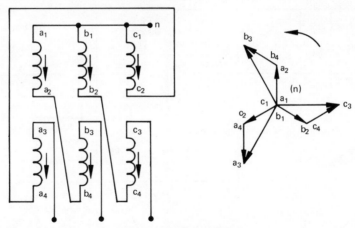

Fig. 5.24   Zig-zag connection and its phasor diagram

ahead of b_1-b_2 and b_3-b_4, which in turn are $120°$ ahead of voltages across c_1-c_2 and c_3-c_4.

Thus voltage phasors a_1-a_2, b_1-b_2 and c_1-c_2 form an ordinary star since the terminals a_1, b_1 and c_1 are joined together to form a neutral point. Furthermore since the terminal a_2 is connected to b_4 then the voltage phasor b_3-b_4 must be reversed and joined to a_1-a_2. The total phase winding voltage is thus given by the phasor n-b_3. Phasors n-a_3 and n-c_3 are obtained in a similar manner.

The zig-zag connection is commonly used in conjunction with rectifiers, or for earthing three wire systems.

5.18   Standard three-phase transformer connection

There are many combinations in which the primary and secondary windings of transformers employed in 3-phase systems may be connected, and therefore a uniform method of grouping these is necessary.

Twelve main connections from British Standard 171 are reproduced in Fig. 5.25. These are arranged in four main groups according to the phase difference between the corresponding LINE voltages on the H.V. and L.V. sides of a given transformer. The phase difference is the angle by which the L.V. line voltage lags the H.V. line voltage and is measured in units of $30°$ in a clockwise direction.

Thus a connection designated by 41 Dy 11 gives the following information.

4 – indicates fourth main group
1 – indicates first connection within the group
D – indicates that H.V. is connected in delta
y – indicates that L.V. is connected in star (a y configuration).

and $11 = 11 \times 30° = 330°$ shows that L.V. line voltage lags H.V. line voltage by $330°$ measured from H.V. phasor in a *clockwise* direction.

Phase Dis-place-ment.	Main Group No.	Vector Group Ref. No. & Symbol.	Marking of Line Terminals and Vector Diagram of Induced Voltages.		Winding Connections and Relative Position of Terminals.
			H.V. Winding.	L.V. Winding.	
Col. 1	2	3	4	5	6
0°	1	11 Yy 0	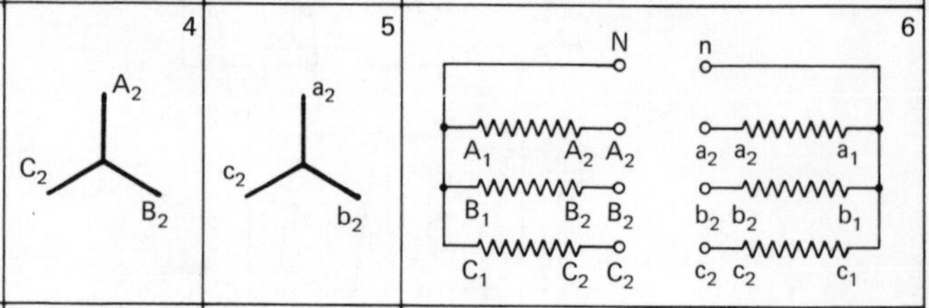		
		12 Dd 0	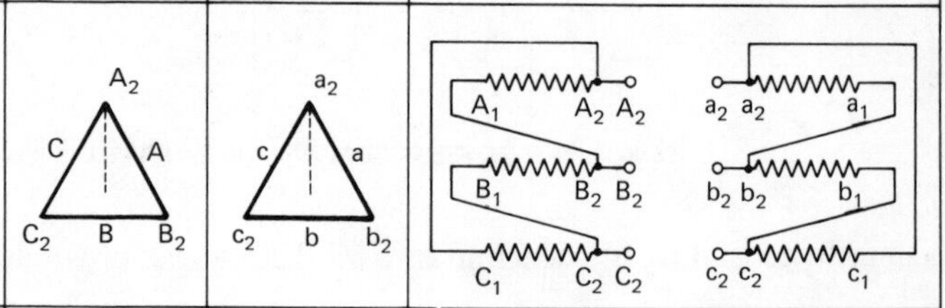		
		13 Dz 0	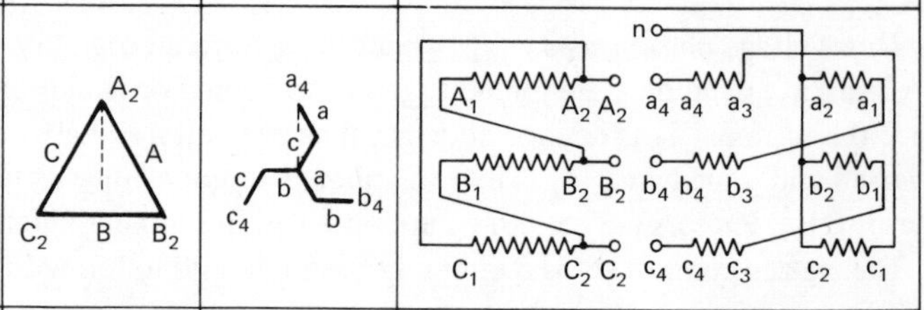		
180°	2	21 Yy 6	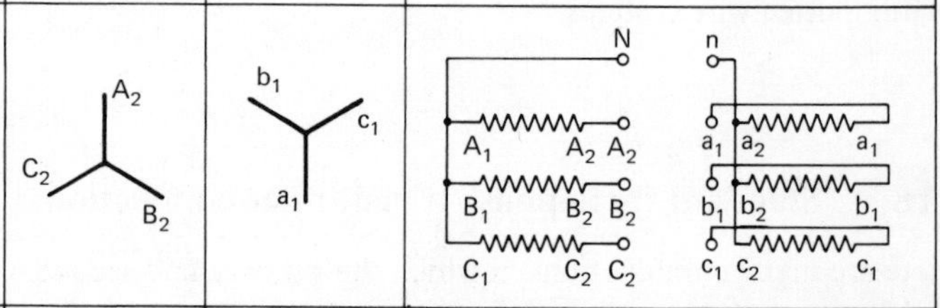		
		22 Dd 6	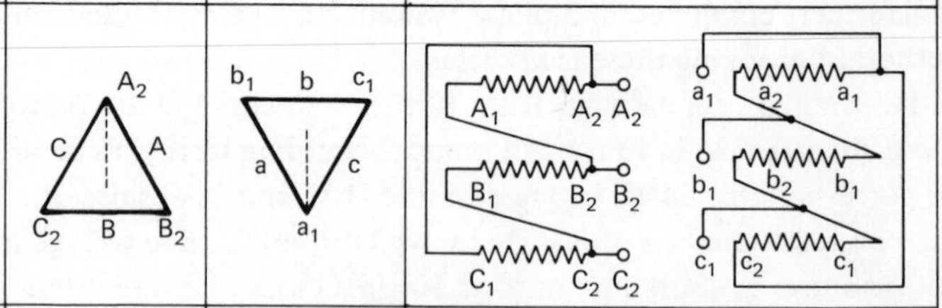		
		23 Dz 6	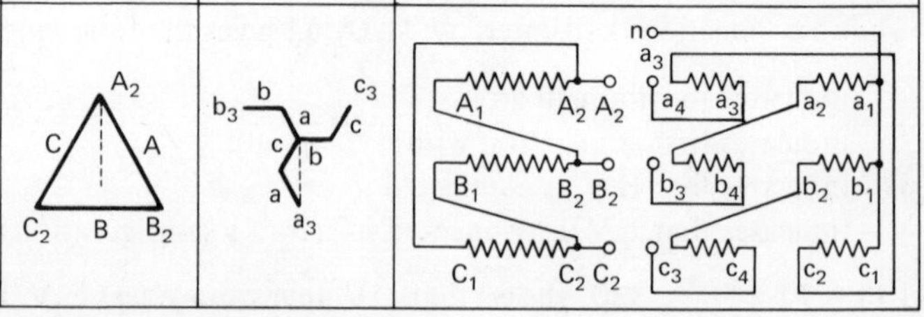		

Fig. 5.25   Standard connections for three-phase transformers

Phase Dis-place-ment.	Main Group No.	Vector Group Ref. No. & Symbol.	Marking of Line Terminals and Vector Diagram of Induced Voltages.		Winding Connections and Relative Position of Terminals.
7	8	9	H.V. Winding. 10	L.V. Winding. 11	12
−30°	3	31 Dy 1			
		32 Yd 1			
		33 Yz 1			
+30°	4	41 Dy 11			
		42 Yd 11			
		43 Yz 11			

QUESTION 8. Draw a complete phasor diagram for a 3-phase transformer which is connected according to the designation 3I Dy I.

ANSWER. The connections of H.V. and L.V. windings are first drawn according to B.S. 171, and the voltage arrows inserted pointing towards terminals with higher subscripts. The H.V. voltage arrows are then placed between the supply lines in such a way that they oppose the winding voltages, because V_{RY} must be equal to $V_{B_1 B_2}$ as demanded by Kirchhoff's voltage law.

Similarly $V_{YB} = V_{C_1 C_2}$ and $V_{BR} = V_{A_1 A_2}$

The L.V. voltage arrows must therefore follow similar sequence from line R to line Y, Y to B and B to R.

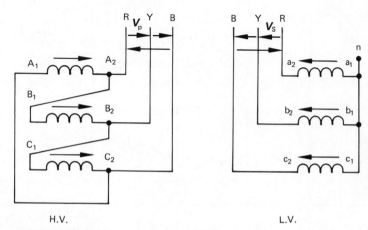

H.V.   L.V.

Fig. 5.26   Connection diagram for Question 8

Assuming the supply sequence to be R, Y, B the supply voltage phasors will form a star configuration as shown. Joining points marked R, Y and B as per line voltage arrows, the phasor diagram for the H.V. side is obtained.

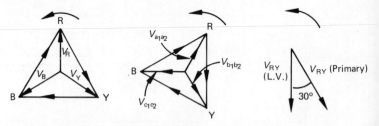

Fig. 5.27   Phasor diagrams for Question 8

Phase voltage phasors for L.V. windings are now drawn parallel to the corresponding H.V. winding phasors, since the H.V. and L.V. windings are placed on the same limb for each phase and therefore their e.m.f.'s must be in phase.

L.V. line voltages are obtained by joining the tips of the phasors forming the star configuration. Comparing phasor V_{RY} on the H.V. with V_{RY} phasor on the L.V. side, it is seen that the L.V. line voltage lags H.V. line voltage by $30°$.

It must be noted that voltage drops in resistances and leakage reactances are neglected in this exercise.

5.19   Transformer circuit for three- to six-phase conversion

The most common method of converting 3 to 6 phases is to use a 3-phase transformer whose primaries are connected in delta and the secondaries (two per phase) joined in double star (Dyy). The circuit is shown in Fig. 5.28(a) and the phasor diagram in Fig. 5.28(b). The latter is obtained in the usual way using the arrow convention of this book.

This method is often used to supply rectifiers for converting alternating current to direct current because the increase of phases from 3 to 6 improves the performance of the rectifier units.

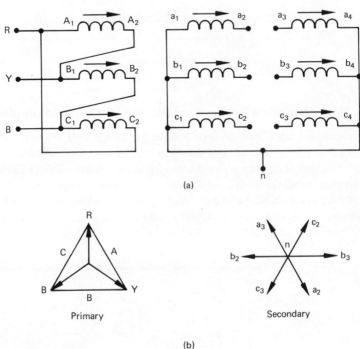

(a)

Primary   Secondary

(b)

Fig. 5.28   Three/six phase conversion using Dyy transformer connection

5.20   Auto-transformers in three-phase supplies

Three single-phase auto-transformers may be arranged to form a 3-phase bank or a single 3-phase unit may be constructed on the common magnetic circuit. In the latter case a 3-limb iron core is used as shown in Fig. 5.23(a).

Unlike the transformers with separate primary and secondary windings the auto-transformer windings are invariably connected in star. The circuit diagram is shown in Fig. 5.29.

They find application as a 3-phase variable supply source, as starters for induction motors or as interconnectors between parts of the national grid. In the last case their size is in the region of hundreds of MVA's and they join 132 kV to 275 kV lines and 275 kV to 400 kV lines.

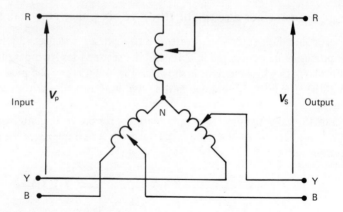

Fig. 5.29   Three-phase auto-transformer circuit

5.21   Moving coil voltage regulators in three-phase supplies

To construct a M.C. voltage regulator for work on a 3-phase system, three single phase units described previously and shown in Fig. 5.6 are mounted on a common base. The movable short-circuited coils are arranged to be driven by a single mechanical device. The main windings are invariably connected in star and the complete circuit diagram is shown in Fig. 5.30. Their main applications are in transmission and distribution systems to maintain constant level of voltage for different load conditions. They are built for voltages of up to 33 kV and from few kVA to over 10 MVA in rating.

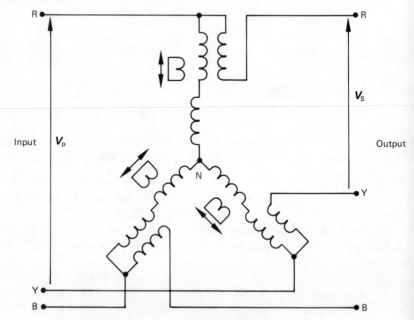

Fig. 5.30   Three-phase M.C. voltage regulator circuit

5.22   Parallel operation of transformers

As the demand on the electrical power system grows, it is frequently necessary to increase the rating of the supply transformers. Rather than take out an old unit and replace it by a new and larger one, it is often more economical to provide an extra transformer and connect it in parallel with the existing one.

5.23   Single-phase transformers in parallel

The two single phase transformers are said to be in parallel when their primary windings are connected to the *same* supply *and* their secondary windings supply the *same* load. Figure 5.31, shows the circuit diagram of two transformers 'A' and 'B' working in parallel.

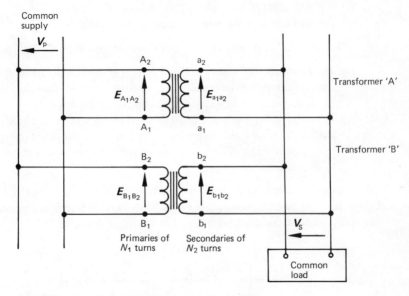

Fig. 5.31   Two single-phase transformers in parallel

Applying Kirchhoff's voltage law to primary and secondary circuits, two equations are obtained:

for primary side

$$V_p = E_{A_1 A_2} = E_{B_1 B_2} \tag{5.29}$$

and for secondary side

$$V_s = E_{a_1 a_2} = E_{b_1 b_2} \tag{5.30}$$

It is clear from their inspection that $E_{a_1 a_2}$ must be in phase with $E_{b_1 b_2}$ because V_p is common and equal to $E_{A_1 A_2}$ and $E_{B_1 B_2}$. Thus the transformer terminals must be correctly marked and like terminals joined together. If this is not done a short circuit will result.

Dividing equation (5.29) by (5.30).

$$\frac{V_p}{V_s} = \frac{E_{A_1 A_2}}{E_{a_1 a_2}} = \frac{E_{B_1 B_2}}{E_{b_1 b_2}} = \frac{N_1}{N_2} \tag{5.31}$$

This shows clearly that the turns ratio of both transformers must also be equal, otherwise the circulating current will flow between the two transformers, in order to equalize the voltages in each loop.

This current will reduce the capacity of the two transformers to supply their full loads to the external circuit. In practice it is impossible to manufacture two completely identical transformers and so there will always be a small unbalance current. However the units will operate satisfactorily, as long as its value is kept small.

Therefore for satisfactory operation, two main conditions are necessary:

     (i) The polarities of the transformers must be the same

and   (ii) the turn ratios of the transformers must be equal.

For efficient operation two further conditions are desirable:

     (i) The potential differences at full load across the transformers' internal impedances *should* be equal. This conditions ensures that the load sharing between them is according to the rating of each unit;

and   (ii) The ratio of their winding resistances to reactances should be equal for both units. This conditions ensures that both transformers operate at the same power factor, thus sharing power and reactive volt-amperes according to their rating.

Assuming that all the conditions are satisfied and that the terminals are correctly labelled, then the two transformers are paralleled by including first a voltmeter between terminal b_1, and the common secondary bus-bar '2' as shown in Fig. 5.22.

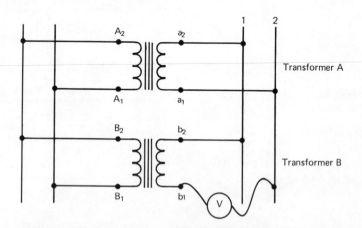

Fig. 5.32 **Paralleling single-phase transformers**

If the voltmeter reads zero then the terminals of like polarities are joined correctly. The instrument may now be taken out of the circuit and b_1 connected to bus-bar '2'.

The voltmeter reading for reversed polarities would be twice the voltage of each transformer secondary,

i.e. $2 \times V_{a_1 a_2}$ or $2 \times V_{b_1 b_2}$.

5.24   Three-phase transformers in parallel

In Fig. 5.25 four main groups of connections were shown for transformers operating in three-phase supply systems.

Each group contains three standard connections which have a common phase difference between the corresponding line voltages on the primary and the secondary sides. It is therefore only the transformers within each group which can be connected in parallel with each other, otherwise even if the line voltages for both are equal and correct terminals are joined together, there would be a resultant voltage due to unequal phase shift between them.

Thus a further essential condition applicable to 3-phase transformers is that the phase shift introduced by their connections must be the same for both units.

The other two conditions already required for single phase transformers must also be satisfied, but may be restated as follows:

(i) The polarities must be the same for all three corresponding phases so that their phase sequence is the same;

and (ii) The turns ratio must be such that line voltages on both sides of the units working in parallel must be equal.

The latter condition is illustrated in the following example:

QUESTION 9. Calculate turns ratio between primary and secondary phase windings for transformers 31Dy1 and 32Yd1, which are to operate in parallel and to step 3-phase 11 000-V supply down to 440 V.

ANSWER.

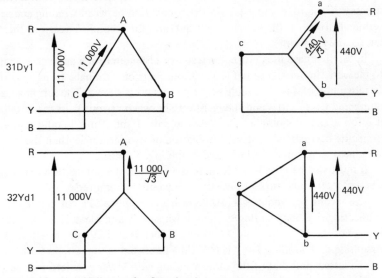

Fig. 5.33   Circuit diagram for Question 9

$$\text{turn's ratio } (31\text{Dy}1) = \frac{11\ 000}{440/\sqrt{3}} = \frac{1100\sqrt{3}}{440} = \underline{43\cdot3}$$

$$\text{turn's ratio } (32\text{Yd}1) = \frac{11\ 000\sqrt{3}}{440} = \frac{11\ 000}{440\sqrt{3}} = \underline{14\cdot4}$$

Finally the two desirable conditions for single phase transformers ought also to be satisfied for efficient operation of the 3-phase units in parallel.

The actual paralleling of two 3-phase transformers is done by first connecting their primaries to the common supply and testing the phase sequence of the secondary voltages by means of a phase sequence meter. If the sequence is the same for both then the secondary circuit is wired as shown in Fig. 5.34.

The two voltmeters V_1 and V_2 must read zero voltage if the essential conditions are satisfied. In this case the instruments are taken out and the secondary connections remade.

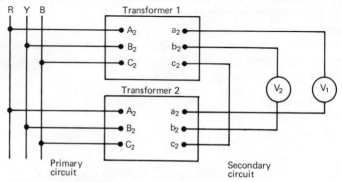

Fig. 5.34   **Paralleling three-phase transformers**

5.25   Labelling transformers with unmarked winding ends

In order to mark the transformer terminals according to B.S. 171 two simple tests are necessary. The first is to establish the connection between winding ends and the terminals to which they are brought out, and the second to assign to them the correct letters and subscripts.

In the first case the continuity tester or ohmmeter such as Avometer may be used to measure the resistance between various pairs of terminals. In Fig. 5.35(a) the terminal box of a two-winding single-phase unit is shown. The terminals are first numbered from 1 to 4 and the resistance between various pairs measured in turn. If it is found that the ohmic value between pairs 1 and 2 and 3 and 4 is low, whereas it is infinite between any other combination of two terminals, then the winding ends are joined to terminals as indicated in the Fig. 5.35(a).

It must also be noted that if there is a break in the connection between the winding end and a terminal, further investigation would be required.

The second test requires a low voltage a.c. supply and a voltmeter with prods. By connecting any one winding to the known supply and measuring the open circuit voltage of the other, it is quickly found, which is an H.V. and which an L.V. winding. Assuming that winding $1 - 2$ is H.V., the next step is to mark its terminals with capital letters and that of the other winding with lower case letters. The subscript numbers are placed arbitrarily as shown in Fig. 5.35(b).

The British Standard requires an e.m.f. of an H.V. winding to act from A1 to A2 at the same instant when the e.m.f. in the secondary winding acts from a_1 to a_2. The transformer windings are first connected as in Fig. 5.35(c), and $V_{A_1 A_2}$, $V_{a_1 a_2}$ and V_{supply} are measured.

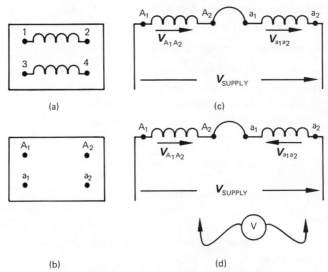

Fig. 5.35   Labelling transformer terminals

The markings are correct if the voltage $V_{A_1 A_2}$ is *less* than the applied voltage V_{supply} because from the circuit it is seen that $V_{supply} \simeq V_{A_1 A_2} + V_{a_1 a_2}$.

If the markings are incorrect the winding voltage $V_{A_1 A_2}$ will be *greater* than the supply voltage, because from Fig. 5.35(d) it is clear that $V_{supply} \simeq V_{A_1 A_2} - V_{a_1 a_2}$.

It is then necessary to interchange the letter subscripts on the L.V. winding and repeat the test.

Care must be taken to supply the transformer with low voltage because in case of the incorrect labelling quite a large current may be drawn by the unit.

SUMMARY

1. Basic transformer consists of two electrical circuits linked by a common magnetic circuit.

2. Transformer changes voltage and current but *does not* transform one kind of energy into another.

3. Transformers operate either as constant voltage or constant current devices.

4. Four types of practical transformers are in general use. These are

      (i) two-winding voltage transformer
      (ii) two-winding current transformer
      (iii) auto-transformer
and   (iv) moving coil voltage regulator

5. Ideal transformer off load behaves like a pure inductive coil, so that current through it *lags* the applied voltage by $90°$.

6. Primary winding e.m.f. $E_p = 4 \cdot 44 \, f N_p \Phi_m$ (V)   (5.3)

Secondary winding e.m.f. $E_s = 4 \cdot 44 \, f N_s \Phi_m$ (V)   (5.4)

where f is the supply frequency in Hz.

$\quad\quad\quad\Phi_m$ is the maximum flux density in Wb.

$\quad\quad\quad N_p$ is the number of turns on the primary winding.

and N_s is the number of turns on the secondary winding.

7. The ampere-turns in a transformer on load obey the equation:

$$I_p N_p - I_s N_s = I_0 N_p$$ (5.6)

where I_p is the primary load current in amperes

$\quad\quad\quad I_s$ is the secondary load current in amperes

and I_0 is a 'no load' primary current in amperes

$\quad$ Since I_0 is small, therefore approximately

$$I_p \simeq \frac{N_s}{N_p} I_s$$ (5.11)

8. In practical voltage transformer the secondary output voltage is reduced by resistances and leakage reactances, whilst the core loss reduces the angle of lag of I_0 to less than $90°$.

9. The following are the differences between voltage and current transformers.

Voltage Transformer
Voltages remain nearly constant
Currents vary with the load
Short-circuit current very high.

Current Transformer
Currents remain nearly constant
Voltages vary with the load
Open circuit voltage very high.

10. The auto-transformer works in a similar manner to a two-winding voltage transformer and its voltages and currents depend on the turns ratio.

11. The operation of the moving coil voltage regulator depends on the alteration of the impedance of the two primary windings by means of a short-circuited movable coil.

12. Transformer losses are due to

$\quad\quad$ (i) core loss $= V_p I_0 \cos \theta_0$ (W)   (5.21)

and   (ii) resistive loss $= I_s^2 \, R_T$ (W)   (5.20)

where V_p is the voltage applied to the primary, $\cos \theta_0$ is the no load power factor. R_T is the resistance of both windings transferred to secondary and I_s is the secondary load current.

13. Efficiency of the transformer can be estimated by performing two tests

$\quad\quad$ (i) Open circuit test

and   (ii) Short circuit test.

The first test gives

$\quad\quad$ (a) core loss at the rated voltage

$\quad\quad$ (b) the value of no-load current I_0

and   (c) data to calculate no-load power factor.

The second test provides the data from which the following can be calculated:

     (a) total winding resistance
     (b) total winding leakage reactance
and (c) resistive loss at any load current.

14.
$$\text{Transformer efficiency} = 1 - \frac{I_s^2 R_T + P_c}{V_s I_s \cos\theta_s + I_s^2 R_T + P_c} \text{ p.u.} \qquad (5.25)$$

where $V_s I_s \cos\theta_s$ = transformer output in watts

and P_c = core loss in watts.

The maximum efficiency occurs when

                        core loss = resistive loss.

15. The output characteristics are graphs which relate:

Secondary terminal voltage V_s to load current I_s when a transformer works as a constant voltage device or Error (Phase angle and Ratio) to load current I_s when the transformer works as a constant current device.

16. Per unit regulation for voltage transformer is given by:
$$\frac{V_s \text{ (no load)} - V_s \text{ (on load)}}{V_s \text{ (no load)}} \text{ p.u.} \qquad (5.27)$$

17. % Ratio (current) error in current transformer is specified as:
$$\% \text{ R.E. at } I_p = \frac{K_n I_s - I_p}{I_p} \times 100\% \qquad (5.28)$$

where K_n is nominal current ratio.
Phase angle error in current transformers is given by the angle in degrees and minutes by which I_s *leads* I_p.

18. Two-winding voltage transformers can be used in 3-phase supply system either in

     (i) banks of 3 single phase units
or (ii) as a single 3-phase unit.

19. Three-phase transformer consists of two separate windings per phase with a common magnetic circuit for all phases.

20. There are many possible combinations of interconnecting three phases of transformers and therefore standard methods of doing so have been established. These are given by B.S. 171.

21. Transformers with many windings can be used to convert 3-phase supply to M-phase supply where M may be any whole number from 1 to 24. Usual conversion is from 3 to 6 phases for use with rectifiers.

22. Both auto-transformers and moving coil voltage regulators are employed in 3-phase supply systems using 3-single phase units. Three phase units are also built.

23. Transformers often operate in parallel. This means that both their primaries *and* secondaries must be joined together.

24. Essential conditions for satisfactory parallel operation of transformers are:
    (i) Their polarities must be the same.
    (ii) Their turns ratios must be equal.
    (iii) They must belong to the same 'phase' group and their phase sequence must
        be the same.

Condition (iii) applies to 3-phase transformers only.

EXERCISES

1. (a) Sketch a phasor diagram for a single-phase transformer on no-load and explain the action of the transformer under these conditions. Include in the description each of the phasor quantities on the diagram.

(b) A 660/220-V single-phase transformer takes a no-load current of 2 A at a power factor of 0·225 lagging. The transformer supplies a load of 30 A at a power factor of (i) 0·9 lagging and (ii) 0·95 leading. Neglecting the effect of winding impedance, determine, using a graphical construction, the current taken from the supply and the primary phase angle in each case.

Answer: 11·4 A   34° 10° lag
                10 A   6° 30′ lead

2. A single-phase transformer has a turns ratio of 144/432 and operates at a maximum flux of 7·5 mWb at 50 Hz. When on no-load the transformer takes 0·24 kVA, at a power factor of 0·259 lagging, from the supply. If the transformer supplies a secondary load of 1·2 kVA at a power factor of 0·8 lagging, determine, with the aid of a scale phasor diagram:

    (i) the core loss current,
   (ii) the magnetising current,
  (iii) the primary current,
  (iv) the primary power factor.

Answer: 0·26 A,   0·97 A.
                5·85 A,   0·73 lag.

3. A small substation has a single-phase 6600-V/240-V transformer supplying four feeders which take the following loads:

   10 kW   at 0·8 p.f. lag
   50 A   at 0·7 p.f. lag
    5 kW   at unity p.f.
   8 kVA   at 0·6 p.f. lead

Determine the primary current and the power factor which the transformer takes from the 6600-V system. (Neglect losses in transformer.)

Answer: 4·53 A.   0·94 lag.   (C. & G.L.I.)

4. A 3300/250-V single-phase transformer has a full-load output of 100 kVA at a frequency of 50 Hz. When the transformer is connected to a 3300-V supply with the secondary winding open circuited, the supply current is found to be 3 A at a power factor of 0·5 lagging. Find:

   (i) the power taken by the transformer and the volt-amperes required to drive the operating flux,
  (ii) the primary current and power factor if the transformer is supplying full load at a power factor of 0·9 lagging.

Neglect the effects of winding resistance and leakage reactance.

Answer: 4950 W,   8572 VA,   32·8 A,   0·876 lag.

5. Explain why the useful flux of a single-phase transformer remains almost constant between no-load and full load.

A single-phase 6600/400 V, 50 Hz transformer takes a no-load current of 0·7 A at a power factor of 0·24 lagging. If the secondary supplies a current of 120 A at a power factor of 0·8 lagging estimate the current taken by the primary by calculation or construction.

Answer: 7·7 A.   (C. & G.L.I.)

6. Open-circuit and short-circuit tests were carried out on a 3-kVA single-phase transformer and produced results as follows:

Open-circuit, $I_p = 1 \cdot 2$ A; $V_p = 200$ V; $W_p = 24$ W.
Short-circuit, $I_p = 15$ A; $V_p = 6 \cdot 4$ V; $W_p = 28$ W.

From the results of these tests determine:

(a) the no-load current and power factor,
(b) the iron losses of the transformer at normal frequency and supply p.d.
(c) the full-load copper losses,
(d) the efficiency of the transformer at (i) half full load, and (ii) full load, at a power factor of $0 \cdot 8$ lagging in each case,
(e) the maximum efficiency of the transformer and the fraction of full load at which it occurs, under the above conditions.

Answer: $1 \cdot 2$ A, $0 \cdot 1$ lag, 24 W, 28 W, 0·975, 0·978, 0·979, 0·926.

7. (a) Describe, with the aid of connection diagrams, the procedure necessary to carry out 'open circuit' and 'short circuit' tests on a transformer. Explain how it is possible to use the results of such tests to determine the efficiency of the transformer at any load.

(b) The maximum efficiency of a 24 kVA single-phase transformer occurs at $0 \cdot 9$ of full-load. If the efficiency of the transformer is found to be 96% at full-load, determine:

(i) the copper loss at full-load,
(ii) the iron loss.

Ignore the effects of reactance of the windings and assume a power factor of $0 \cdot 8$ lagging at all loads.

Answer: 442 W,   358 W.

8. Draw clearly labelled diagrams for the following methods of connecting the windings of three-phase transformers:

(a) star/star with tertiary,
(b) delta/star,
(c) star/interconnected-star (zig-zag).

Show, by phasor diagrams, the relationships between the input and output voltages for each of the above.   (C. & G.L.I.)

9. (a) What conditions must be satisfied in order that two transformers shall operate satisfactorily in parallel?

(b) The supply to a factory is generated by the power station generator at 11 000 V, stepped up to 132 000 V by a mesh-star transformer and fed to the grid system. From the grid the supply passes through a 132/11 kV star/star transformer with a tertiary winding to the primary distribution network from which the factory is fed through an 11/6·6 kV mesh/star works transformer. Sketch the phasor diagram showing the relationship between the line voltages at the various stages and state the phase displacement between the voltage at the generator terminals and that at the works transformer secondary terminals.   (C. & G.L.I.)

Answer: Transformer secondary line voltage lags by 60° the generator line voltage.

10. Describe, giving connection and phasor diagrams, two ways of obtaining a 6-phase supply from a 3-phase source.

Show with the aid of suitably labelled winding diagrams how three single-phase transformers, each having a 240-V primary and two 200-V secondaries, can be connected up to give a 6-phase output from a 415-V 3-phase supply.

Sketch a phasor diagram and calculate the voltage between adjacent phases of the 6-phase voltage.   (C. & G.L.I.)

Answer: 200 V.

11. What are the conditions which must be satisfied in order that the two three-phase transformers may operate satisfactorily in parallel? Describe how you would determine, by tests, the correct connections when installing a three-phase transformer, whose terminals were unmarked, to operate in parallel with an existing one.   (C. & G.L.I.)

12. Show, by suitably labelled winding diagrams, how three single-phase transformers, each having a 240-V primary and two 250-V secondaries, can be connected up to give a six-phase output from a 415-V, 3-phase supply.

   Sketch the voltage phasor diagram and hence obtain the open circuit voltage between adjacent phases of the 6-phase voltage.

   If one of the secondaries were reversed what would be the voltage between it and the adjacent phase?   (C. & G.L.I.)

Answer: 250 V,   415 V.

Windings, currents and e.m.f.s in rotating machines

In the study of electrical machines the energy conversion region (described in Chapter 3), is of the utmost importance. The region illustrated in Fig. 3.7 shows clearly the magnetic field as an agent through which the mechanical power is converted to electrical power and vice versa. Furthermore, the pattern *of the magnetic field of a rotating machine determines its type. The field pattern, in turn, is produced by* currents *flowing in the* windings *mounted on the stator and the rotor of a given machine.*

In this chapter therefore, the windings employed in rotating machines are considered in greater detail, as well as the magnetic flux patterns and e.m.f. waveforms which they generate. The nature of direct and alternating currents is described in Chapter 2 'Review of circuit theory'.

6.1 Classification of windings

The windings used in the construction of electrical machines are either *concentrated* or *distributed.*

The *concentrated* windings are invariably of the *coil* type. The coils which form a complete winding consist of turns tightly wound on a central core, each turn being adjacent to the next. The *distributed* windings are placed in slots which are cut around a part of, or the whole of, the circumference of cylindrical rotors and stators. The distributed windings may either be:

    (i) coil windings
   (ii) cage windings
  (iii) phase windings
or (iv) commutator windings.

6.2 Concentrated winding – coil type

The winding consists of coils wound on salient poles and connected together in series or in parallel to form a single circuit. Figure 6.1(a) shows a portion of the stator whereas Fig. 6.1(b) illustrates a section of the rotor – both with two projecting poles carrying a concentrated coil winding. The end connections of the whole winding are brought out to fixed terminals if the winding is placed on the stator, or to sliprings if the winding is arranged on the rotor.

It is to be noted that adjacent poles are usually of alternate polarity North–South–North again, and so on. The stators of this type are employed in direct current machines and small synchronous machines, whereas the rotors are used in large, slow speed, synchronous machines. The e.m.f. of rotation is NOT normally induced in the concentrated coil windings, and their purpose is to provide the main magnetic field in the air-gap of a given machine.

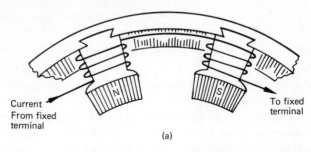

Current
From fixed
terminal

To fixed
terminal

(a)

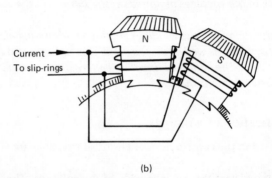

Current
To slip-rings

(b)

Fig. 6.1   Salient poles with a concentrated coil winding (a) Stator (b) Rotor

6.3   Slip-ring and brush gear

A device which is used to transfer the current between a continuously rotating member of the machine (rotor) and its stationary part (stator) is shown in Fig. 6.2.

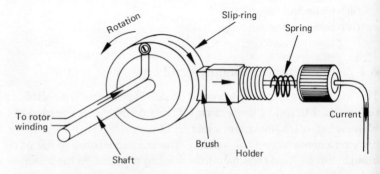

Fig. 6.2   Slip-ring and brush gear

It consists of a conducting ring mounted on the rotor shaft and insulated from it. The end connection of a rotor winding is joined to the ring, which revolves with the rotor.

The carbon brush, supported in a holder attached to the stator, is lightly pressed by the spring to the slip-ring surface. An external circuit is connected to the conducting wire which is embedded in the carbon brush.

The current flows from the rotor winding through the slip-ring, then transfers to the stationary brush and through its connection into an external circuit.

The device just described does not alter the nature of the current, it merely enables it to flow between the supply and the rotor of a machine.

6.4   Distributed windings – introduction

The inner surface of a cylindrical stator, and the outer surface of a round rotor, have the slots cut out in them as shown in Fig. 6.3(a) and (b).

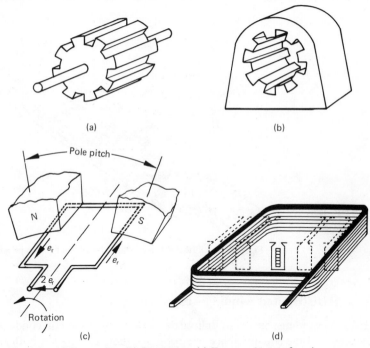

(a)   (b)

(c)   (d)

Fig. 6.3   (a) Rotor slots (b) Stator slots (c) Two conductors forming one turn (d) A single coil composed of six turns.

The conductors are placed in the slots, each pair forming one turn (Fig. 6.3(c)). The conductors are approximately one pole pitch apart, a pole pitch being the distance from the centre of the pole of one polarity (North) to the centre of the adjacent pole of opposite polarity (South). The reason for this spacing is that the e.m.f.'s must add up around the turn, and therefore, must have opposite directions in the two conductors, which form the sides of each turn.

This condition is obtained only when one conductor is under the influence of a 'North' pole, whilst the other under the influence of a 'South' pole for the same direction of rotation. (See Chapter 3.)

The turns joined together as shown in Fig. 6.3(d) form a single coil. Such coils are then connected together in a variety of ways to obtain different types of distributed windings.

6.5   Distributed winding – coil type

In cases when a rotor does not have projecting (salient) poles, the coil-type winding is distributed in the slots cut on its cylindrical surface. Figure 6.4 illustrates this type of a rotor, used in high speed synchronous machines.

The distributed coil winding like the concentrated coil winding, does not normally have an e.m.f. of rotation induced in it, and only serves to provide the main magnetic field for the machine.

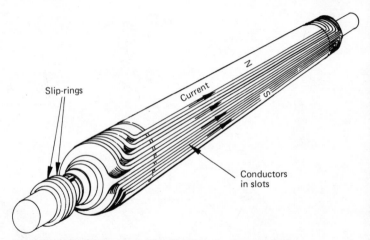

Fig. 6.4   Cylindrical rotor with a distributed coil type winding

6.6   Distributed winding – cage type

The cage winding consists of uninsulated copper or aluminium conductors placed in slots around the circumference of a rotor and short-circuited at both ends by means of conducting rings.

The rotor with a cage winding is illustrated in Fig. 6.5(a). The winding may be thought of as a series of turns, each made up of two conductors spaced one pole pitch apart and connected together at both ends by portions of the short-circuiting rings (Fig. 6.5(b)).

Thus each turn forms a short-circuit for current produced by rotational e.m.f.s induced in its pair of conductors. The currents which flow in this type of winding are due to electro-magnetic induction and are not lead into it from an external electrical source.

Cage windings are used in induction machines and in synchronous machines as damper grids.

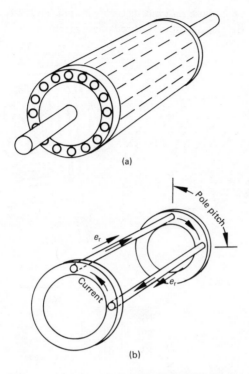

Fig. 6.5   (a) Rotor with a cage winding (b) Conductors forming a turn.

5.7   Distributed winding – phase type

The phase winding has separate external connections, much as a concentrated coil winding. The current enters through one fixed terminal (or slipring) and emerges from the other terminal (or slipring), having passed through the whole winding.

The winding itself consists of a number of coils (Fig. 6.3(d)) which are joined together in series. The coils can be placed in the slots either:

(i) in two layers (double layer winding)
or (ii) in single layer (single layer winding).

In the former case, illustrated in Fig. 6.6(a) all the coils are of the same size and one side of each coil lies at the bottom of a slot, whilst the other is at the top of another slot, approximately a pole pitch away.

Each slot therefore, accommodates two coil sides, belonging to two different coils and insulated lightly from each other. The cross-section in Fig. 6.6(b) shows the two 'layers' in one slot.

In case of a single layer arrangement each slot contains only one coil side and therefore the coils must be of different spans. These coils are shown in Fig. 6.6(c).

A Stator partially wound with a phase winding in a double layer arrangement is illustrated in Fig. 6.7(a). A Stator shown in Fig. 6.7(b) carries a partially completed single layer three phase winding.

It must be noted that each phase requires its own winding. Thus a two phase machine needs two separate phase windings, three phase machine, three windings and so on.

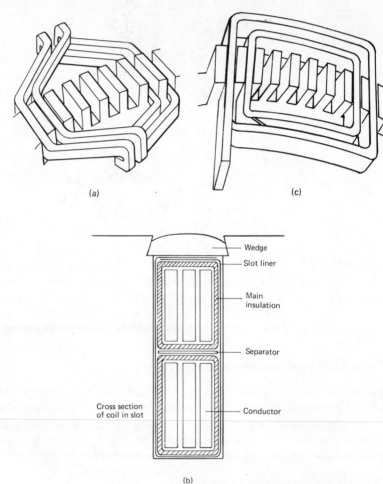

(a)   (c)

Wedge

Slot liner

Main insulation

Separator

Cross section of coil in slot   Conductor

(b)

Fig. 6.6   **(a) Double layer arrangement of two coils in slots (b) Cross-section of a slot containing two coil sides (c) Single layer arrangement of two coils in slots**

6.8   Distributed winding – commutator type

The winding consists of a number of identical coils as shown in Fig. 6.3(d) and shaped as in Fig. 6.6(a), which are placed in the slots around the circumference of a rotor. Each slot accommodates two coil sides, belonging to two different coils as in the case of a double layer phase winding.

Fig. 6.7   (a) Stator with double layer phase winding (b) Stator with single
layer three-phase winding. (*Laurence, Scott and Electromotors
Ltd.*)

The coils are joined together in series, and each junction of all the coil pairs is then
connected to the commutator segment in front of the rotor.

There are two basic ways of connecting coils together and to a commutator; the
one is simplex lap and the other simplex wave. The first is illustrated in Fig. 6.8(a)
and the second in Fig. 6.8(b). In each case only two single turn coils are shown, made
up of conductors of rectangular cross-section and with the insulation omitted for
clarity.

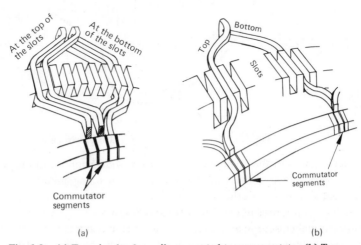

(a)   (b)

Fig. 6.8   (a) Two simplex lap coils connected to a commutator (b) Two
simplex wave coils connected to a commutator

In either case the winding, thus formed, is completely closed and the current is lead into and out of it through brushes resting on the surface of the commutator.

The commutator-type winding is invariably used on rotors and is employed in direct current machines and alternating current commutator machines.

6.9   The commutator

The commutator is a cylinder fitted at one end of the rotor. It is made up of a large number of copper segments. The segments are insulated from each other and from the shaft by thin mica and held together by insulated end rings. Each segment forms the junction between two adjacent rotor coils. The coil wires are soldered to the end of each segment either into milled slots or to flat strip connectors called commutator risers. The commutator of dove-tailed construction is shown in Fig. 6.9.

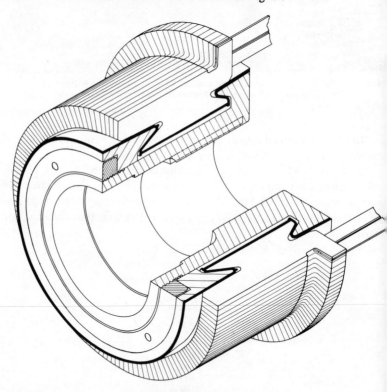

Fig. 6.9   Commutator of dove-tailed construction

The function of the commutator is (i) to provide a connection for current transfer between a stationary brush and the commutator winding, mounted on the rotor and (ii) to act as a 'rectifier', that is, a convertor of an alternating current induced in each conductor as it moves under the North and South poles alternately, into a unidirectional current in an external circuit, i.e., outside the commutator winding. It maybe noted that if the brushes are rotated around the commutator, the device acts as a frequency convertor.

6.10   The rectifier action of the commutator

The 'rectifier' action of the commutator can best be explained by considering a rotor
with 8 slots, carrying a simple lap winding consisting of single turn coils arranged in
two layers in each slot. The rotor is mounted inside a two pole stator. The conductors
are shown in sectional elevation. Their connections to the commutator are indicated
inside and back-end connections outside the rotor circumference. Full lines are used
for conductors lying at the top of the slots and broken lines for those in the bottom.
A single turn coil is formed from two conductors a pole pitch apart. The top conductor
in slot 1 is connected at the back of the rotor to the bottom conductor in slot 5.

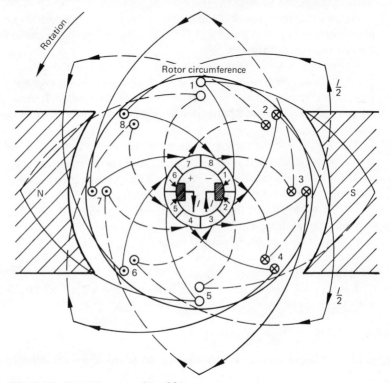

Fig. 6.10 Simple lap wound machine

The ends of this coil are brought out in front of the machine and connected to
two adjacent segments numbered 1 and 2. The next coil lying in slots 2 and 6 is
connected to segments 2 and 3, and so on, until all the coils are joined together. The
coils of the rotor are thus in series and the winding is continuous, that is without a
definite entry or an exit terminal.

If the rotor is driven anti-clockwise, the direction of an e.m.f. of rotation in each
conductor can be found by applying Fleming's Right Hand Rule. The directions of
e.m.f.'s and currents are shown by dots, when they flow out of the paper and by
crosses when they flow *into* the paper. Using these as starting points, current flow in
overhang connections is worked out and is shown in Fig. 6.10 by the arrows. The two
coils in slots 1 and 5 have no current flowing in them since at this instant they occupy

the space between the two poles. Inspection of segments 1 and 2, and 5 and 6, reveals that currents tend to enter the winding through the first pair and leave it through the second. Thus the brushes must be placed at these points, the brush on 5th and 6th segment being positive and that on 1st and 2nd segment being negative. It must be noted that the brushes at that instant short circuit the two inactive coils in slots 1 and 5. The sequence of events in the two coils lying in slots 2 and 6, is now considered as the rotor turns in an anti-clockwise direction.

In Fig. 6.10 the current in the first coil flows from segment 2 through the coil itself towards segment 3, whereas in the second coil the current flows from segment 7 through the coil towards segment 6.

As the rotor turns through 45° the two coils take up the position between the poles, occupied by slots 1 and 5 in Fig. 6.10. The current decreases to zero and the brushes short circuit the two coils.

Further 45° rotation puts the coils in the slots marked 8 and 4. The current in the first coil flows now from segment 3 through the coil towards segment 2, and in the second coil from segment 6 through the coil towards segment 7. Thus the current around each coil reverses its direction as that coil is transferred from one to the other side of the brush.

The direction of the current, however, in the circuit, external to the rotor winding, is still the same, that is from positive brush towards the negative brush.

Thus the commutator and the stationary brush gear convert alternating current in each coil into a direct current outside the winding.

It must be noted that there are two parallel paths inside the considered winding between the two brushes.

In general the number of parallel paths is equal to the number of poles for lap type winding, whereas wave type winding has always two parallel paths irrespective of the number of poles.

A rotor carrying a complete commutator winding is illustrated in Chapter 4, Fig. 4.4(b).

6.11   Magnetomotive force F due to windings – introduction

Having described the main types of windings, it is now necessary to consider the magnetic field patterns created by the currents flowing in them. As stated already, it is through the magnetic field in the air-gap that the electrical power $p_e = e_r i$ is converted to mechanical power $p_m = T\omega$ and vice versa.

The e.m.f. of rotation e_r and the torque T are given by the equations

$$e_r = Bl\omega r \text{ (V)} \tag{3.2}$$

and
$$T = Blir \text{ (Nm)} \tag{3.6}$$

Both these quantities depend on flux density B, which describes the magnetic field at a given point. Hence the *graph of flux density plotted against the circumference of a rotor (or stator)* is used to represent the pattern of the magnetic field existing inside a given machine.

This graph serves in subsequent chapters to predict the waveforms and to calculate the values of e.m.f. and torque for different types of machines. The flux density B measured in teslas is defined as the total magnetic flux Φ in webers passing through an area A square metres, i.e.,

$$B = \frac{\Phi}{A}\,(\mathrm{T}) \qquad\qquad (6.1)$$

The magnetic flux Φ is produced by a magnetomotive force F in amperes, and opposed by the reluctance S in amperes per weber, of a magnetic circuit in which Φ is established. In equation form

$$\Phi = \frac{F}{S}\,(\mathrm{Wb}) \qquad\qquad (6.2)$$

Finally the reluctance S depends on the dimensions and the material from which the magnetic circuit is constructed i.e.,

$$S = \frac{l}{\mu A}\,(\mathrm{A/Wb}) \qquad\qquad (6.3)$$

where l is the length of the magnetic path in metres.

A is the cross-sectional area in square metres.

μ is a constant of the material called the absolute permeability and is measured in henrys per metre.

μ is made up of two parts, i.e.,

$$\mu = \mu_r \mu_o$$

where μ_r is relative permeability and μ_o is the permeability of free space

$$\mu_o = 4\pi \times 10^{-7}\ \mathrm{H/m}$$

Combining equations (6.1), (6.2) and (6.3),

$$B = \frac{\Phi}{A} = \frac{F}{SA} = \frac{F}{lA/\mu A} = \mu\frac{F}{l} = \mu_r\mu_o\frac{F}{l}$$

thus

$$B = \mu_r\mu_o\,\frac{F}{l}\,(\mathrm{T}) \qquad\qquad (6.4)$$

Equation (6.4) may be simplified by observing that in electrical machines the radial length of the air-gap has a very large reluctance compared with that of the magnetic core and therefore requires most of the m.m.f. Hence the effect of a magnetic core can be neglected.

Thus l is reduced to an air-gap length l_g. Furthermore, μ_r for air is very nearly equal to unity, and the expression (6.4) becomes:

$$B = \mu_o\,\frac{F}{l_g} \qquad\qquad (6.5)$$

To obtain the flux density pattern for a given winding, the m.m.f. distribution diagram is first drawn to scale. The m.m.f. F is due to a current I flowing in N turns of that winding so that $F = NI$ amperes. The scale of the diagram is then multiplied by μ_o and divided by the gap length l_g.

6.12 F and B distribution diagrams due to concentrated coil winding

Figure 6.11(a) shows a machine with two salient poles carrying a concentrated coil winding. The machine is redrawn in a developed form and its m.m.f. and flux density curves are shown plotted against its circumference in Fig. 6.11(b).

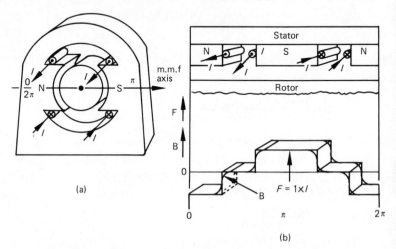

(a)

(b)

Fig. 6.11   F and B distribution diagrams for concentrated coil winding

For simplicity, the winding illustrated has only one turn per pole. If the current flowing is I amperes, then the m.m.f. under each pole is 1 turn x I amperes = I (A). The direction of I around the south pole is anticlockwise and the value of the m.m.f. is constant over the area of the whole turn. It can therefore be represented as a plane raised I units above the datum. Positive value is assigned to this m.m.f. since it produces a magnetic flux in an upward direction (Fig. 6.11(b)).

In the space between the poles the current is not 'circular' and therefore the m.m.f. there is zero.

Under the 'North' pole the current flows in a clockwise direction and therefore the m.m.f. due to it is regarded as negative. Thus the m.m.f. acts from left to right as shown in Fig. 6.11(a) or vertically upwards at the centre of the developed diagram. In Fig. 6.11(b) equation (6.5) shows that the flux density B is directly proportional to the m.m.f. F and inversely proportional to the length l_g of the air-gap. As l_g between the poles is very large and under the poles it is small and of constant value, the shape of B is similar to that of F but with transition portions rounded off because of flux fringing. A graph of B is drawn to coincide with the F curve for easy comparison. Normally the profile of the three dimensional graph is all that need be shown, since its breadth is everywhere the same and equals the axial length of the machine. In the succeeding paragraphs therefore, only two dimensional plots of F and B are shown.

6.13   *F* and *B* distribution diagrams due to distributed windings

CONDUCTORS ONE POLE PITCH APART

Figure 6.12(a) shows a rotor with two conductors spaced one pole pitch apart. The conductors form one turn coil which carries a current of I amperes. The m.m.f. $F = 1 \times I$ and acts from left to right.

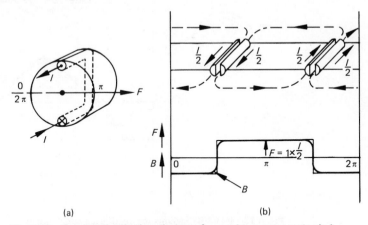

Fig. 6.12   F and B distribution diagrams for conductors one pole pitch apart

To obtain the distribution diagram the two conductors are drawn as shown in Fig. 6.12(b). They may be thought of as forming *two* single turn coils, similar to those wound around the poles in the previous case, but placed close together. One half of each conductor forms the right hand side of a turn to the left of the conductor, whereas the other half forms the left hand side of a turn to the right of it. Each conductor's half carries $I/2$ amperes as shown in Fig. 6.12(b), the dotted lines indicating the turns to which the conductors belong. The m.m.f. diagram therefore consists of two rectangles the height of each being 1 turn $\times I/2$ amperes $= I/2$ (A).

As before the flux density curve has the shape similar to the m.m.f. wave but with the corners rounded off.

By comparing the m.m.f. diagrams of Fig. 6.11(b) and 6.12(b) it is clear that the single turn coil is equivalent to two salient poles.

DISTRIBUTED WINDING – COIL TYPE

The rotor with coil type winding is illustrated in Fig. 6.13(a). It is shown again, but in a developed form together with F and B diagrams in Fig. 6.13(b).

Each 'band' or section, consisting of a number of conductors (five in this case) carries a current in the same direction, the direction in adjacent sections being opposite to each other. Just as in the previous paragraph each conductor is divided into halves, so in this case each band of conductors is halved and the turns formed as indicated by dotted lines.

The resultant m.m.f. wave due to the current of I amperes in each conductor consists of two trapeziums, each with stepped sides.

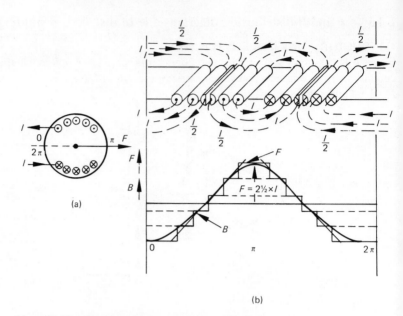

Fig. 6.13   F and B diagrams for distributed coil winding

Every turn adds 1 x *I* amperes on to the previous turn of larger span, the transition being sharp and occurring at each conductor. If the band width consists of an odd number of conductors the middle conductor is halved; if the number is even the centre of the band occurs between the two conductors and no splitting is necessary.

As the number of conductors in each band is increased, the steps on the sides of each trapezium become smaller, so that for a large number of conductors per band it is usual to assume a *straight line* transition shown in Fig. 6.13(b).

Finally, the flux density curve is fitted to the m.m.f. graph by rounding off the intersections of straight lines. It is clear that the curve approximates to a sinusoidal distribution of flux density in the gap. Since this is the usual aim in designing electrical machines, it will be assumed in this book that a distributed coil winding produces sinusoidal flux density. As such, the m.m.f. causing it can be represented by a stationary space phasor *F*, shown directed from left to right in Fig. 6.13(a). The winding shown is equivalent to two salient poles. It should be noted however, that by doubling the number of bands the winding will give four equivalent poles, trebling it six poles, and so on.

DISTRIBUTED WINDING – SINGLE PHASE TYPE

The winding (Fig. 6.14(a)) usually occupies the whole circumference of a rotor or stator and is placed in a large number of slots. Each pole pitch accommodates the conductors which carry current in the same direction. In Fig. 6.14(b) the developed form of the winding is shown illustrating two equivalent poles.

Applying the same method of dividing each band of conductors in the pole pitch in half and forming the turns, the triangular m.m.f. diagram is obtained. The flux density curve *B* when drawn inside the triangular *F* wave may again be assumed to be sinusoidal.

Phase windings usually carry alternating current which increases from zero to a maximum value, decreases to zero, reverses its direction and repeats its growth and decrease pattern. Hence the m.m.f. curves change from zero lines to triangular wave shapes, each triangle increasing to its maximum height and then decreasing. Similarly the flux density distribution waves vary between zero and sinewaves of maximum amplitude. Such waves are said to be pulsating. The dotted lines in Fig. 6.14(b) indicate wave shapes for different instantaneous values of an alternating current flowing in conductors of the winding. In Fig. 6.14(a) the m.m.f. *F* is again represented by a stationary phasor directed from left to right. But unlike the phasor for an m.m.f. of a distributed winding, its length is no longer constant but changes with varying values of the alternating current. As before, this is justified by the fact that the m.m.f. wave *F* although not sinusoidal itself nevertheless produces a sinusoidal flux distribution in the air-gap of a machine.

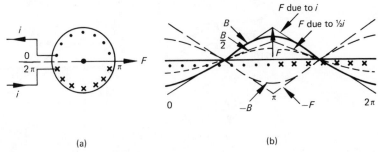

(a)   (b)

Fig. 6.14   F and B diagrams for single phase winding

DISTRIBUTED WINDING – TWO PHASE TYPE

An example of a two phase winding is shown in Fig. 6.15(a). In Fig. 6.15(b) the m.m.f. and flux density waves are drawn at two different instants of time. It must be noted that each phase must have a winding of its own, hence in the illustration the rotor carries two single phase windings interconnected to form a two-phase type. Thus each pole pitch accommodates two bands of conductors, one for each phase, fed from a two phase supply. A two phase supply is that in which an alternating current in phase I leads the current in phase II by the time interval equivalent to an angle of $\pi/2$ radians.

The m.m.f. and flux density waves are drawn at two instants, the time interval between them producing a phase shift of $\pi/2$ radians. The first instant is when the current in the phase I winding is at a maximum and that in phase II is zero. The second instant is when the reverse condition obtains. Inspection of the waves shows that m.m.f.'s are trapezoidal in shape and that flux density plots may be assumed to be sinusoidal. The waves however do not pulsate, i.e., vary in amplitude. Their shape remains approximately constant, but their position with respact to the conductors shifts from right to left. The shift is equal to $\pi/2$ radians, i.e., the same as the time interval between the two instants at which graphs in Fig. 6.15(b) were plotted. The continuous shift produces rotation when the winding is placed on the rotor (or stator). Thus in general a two phase winding fed from a two phase supply generates a sinusoidal flux distribution in the air-gap which is constant in magnitude and rotates at a rate dependant on the frequency of the supply currents. In Fig. 6.15(a) the m.m.f.

in such a winding is again represented by a space phasor, which in this case has constant magnitude and rotates with respect to the surface of the rotor or the stator. The winding shown is equivalent to *two rotating* salient poles.

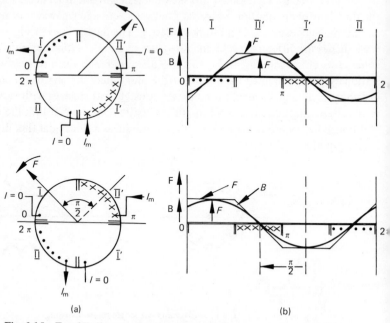

(a)

(b)

Fig. 6.15   F and B diagrams for two-phase winding

DISTRIBUTED WINDING – THREE-PHASE TYPE

A three-phase winding equivalent to two salient poles is considered. It consists of three separate phase-type windings which are connected in STAR. Each winding occupies $\frac{1}{3}$ of the rotor's (or stator's) circumference and its axis is physically displaced from its neighbours by $\frac{2}{3}\pi$ radians (120°). Thus each equivalent pole pitch accommodates three bands each belonging to a different winding.

The first winding is marked RR′ (Fig. 6.16(b)) where the first conductor of R band is regarded as its 'START' and the last conductor of the R′ band as its 'FINISH'. Similarly, the second winding is marked YY′ and the third BB′. It is clear that 'START' conductors of each phase are displaced to each other by $\frac{2}{3}\pi$ radians (120°).

The windings are supplied with three-phase alternating currents, the graphs of which are shown in Fig. 6.16(a). The three currents are out of step with each other by the time interval equivalent to $\frac{2}{3}\pi$ radians (120°), and the current i_R flows in coil RR′, i_Y in coil YY′ and i_B in coil BB′. At the instant of time marked by the line XX′ in Fig. 6.16(a), the instantaneous values of the currents read from the graphs are as follows:

$$i_R = + I_m$$

$$i_Y = -\frac{I_m}{2}$$

$$i_B = -\frac{I_m}{2}$$

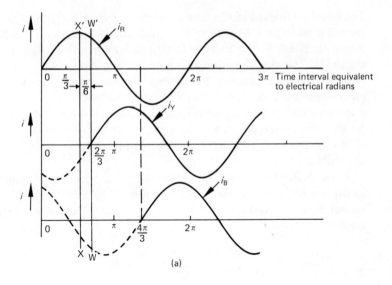

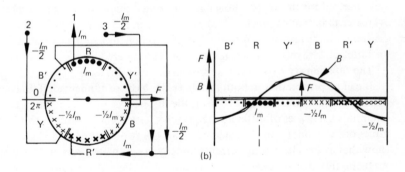

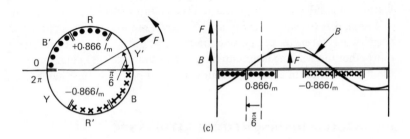

Fig. 6.16   (a) (b) (c) F and B diagrams for three-phase winding

The values are marked in Fig. 6.16(b) where a positive direction of current through the coil is taken to be from R' to R, Y' to Y and B' to B. Thus i_Y and i_B enter through terminals '2' and '3', join together into i_R and return to the supply via terminal marked '1'. It is clear therefore, that no more than three wires connecting the supply to the machine are necessary. Re-drawing the rotor in a developed form and using previous methods, the m.m.f. wave for instant XX' is obtained. It is seen that its shape consists of a trapezium with a triangle superimposed on it. The m.m.f. axis occurs at the central point of the winding, i.e., it acts from left to right when the undeveloped rotor is considered. The flux density curve drawn inside the m.m.f. wave approximates to a sinewave.

In Fig. 6.16(c) the same windings are shown carrying current at an instant marked by the line WW' in Fig. 6.16(a). The time interval between lines XX' and WW' is equivalent to $\pi/6$ radians (30°) and the new instantaneous values of alternating currents are,

$$i_R = + 0{\cdot}866\, I_m$$
$$i_Y = 0$$
$$i_B = - 0{\cdot}866\, I_m$$

This means that the current $0{\cdot}866\, I_m$ now enters the winding BB' and returns to the supply via the winding RR', the winding YY' being inactive at this instant. The m.m.f. wave derived for this instant now has trapezoidal shape, the height of which however is greater than that of previous trapezium. The m.m.f. axis is also no longer at the central point but has moved to the left by $\pi/6$ radians (30°), or rotated $\pi/6$ radians in an anticlockwise direction when the rotor is considered.

The flux density curve again approximates to a sinewave.

It can be proved that in both instances the basic or fundamental sinewave of flux density B is the same in every respect, the differences being accounted for by small-amplitude sinewaves of different frequencies termed harmonics. In this book the effect of harmonics is not considered. Thus the two cases described are sufficient to show that as the alternating currents undergo a complete cycle of 2π radians, the magnetic flux density distribution around the surface of the rotor rotates through one full revolution of 2π radians. In general, therefore, the three-phase winding shown spaced around the circumference of the rotor (or stator) and fed with three-phase alternating currents is equivalent to a two pole system producing a constant magnetic field sinusoidally distributed in the air-gap, which rotates in space with the speed such that the time taken by each revolution is equal to the time of one current cycle.

In Fig. 6.16(b) and (c) the m.m.f. F is again represented by a phasor of constant magnitude rotating in space inside the stator or outside the rotor surface according to whether the winding is mounted on one or the other. By doubling the number of conductor 'bands' in a machine, four equivalent poles can be obtained, and so on.

DISTRIBUTED WINDING – COMMUTATOR TYPE

A commutator winding and its m.m.f. distribution diagram are shown in Fig. 6.17(a) and (b).

The shape of the m.m.f. wave is triangular, just as in the case of a single-phase winding, but unlike it, it remains stationary because the current direction between any two brushes is unchanged. As before the height of each triangle depends on the magnitude of the current and the number of conductors per pole. As explained in paragraph 6.10 the change in current direction occurs at each brush, and hence the m.m.f. axis also occurs at this point.

The flux density curve B is not shown in Fig. 6.17, because the rotors with commutator windings are used in machines with non-uniform air-gaps, i.e., where stators have salient poles.

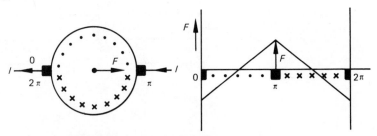

Fig. 6.17 F diagram for commutator winding

6.14 Magnetic field pattern due to two windings

Various examples given in Chapter 4 show that any rotating machine consists of a minimum of two windings; one placed on a stator and the other on a rotor.

When a machine is run unloaded then in general only one of its two windings carries a current and its function is to establish the main magnetic field.

When however, a machine is loaded either as a generator or as a motor, the second winding, usually one of the distributed types, also carries a current and has an e.m.f. of rotation induced in it. The magnetomotive force which is now responsible for producing the magnetic flux density in the machine's air-gap is thus the resultant of the waveforms due to *both* windings. This effect is known as 'armature reaction' and occurs in every machine operating on load. The following examples illustrate the method which enables the resulting field pattern to be obtained under these conditions.

QUESTION 1. A machine consists of a four pole stator and a rotor carrying a communator winding. The particulars are as follows:

STATOR: Each salient pole has 200 turns of concentrated coil winding, carrying a d.c. current of 0·1 A. The length of the pole shoe is equal to $\frac{5}{6}$ of the pole pitch.

ROTOR: Number of slots per pole = 6 spaced at equal intervals.
Number of conductors per slot = 2
Direct current per conductor = 2 A
Brush position midway between the poles

Sketch a developed diagram of the machine and plot to scale:

(i) Magnetomotive force diagram for the stator winding only,
(ii) Magnetomotive force for the rotor winding only,
(iii) Combined magnetomotive force due to currents in both windings.

Assume the machine to be run as a generator, the rotor turning in an anti-clockwise direction.

ANSWER

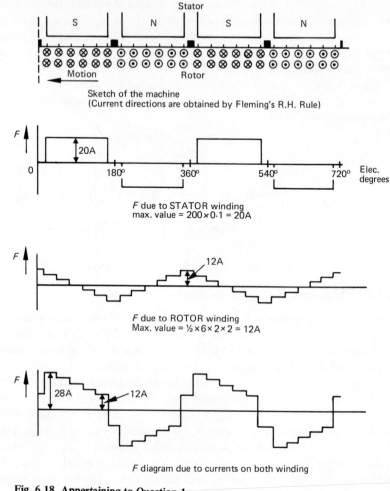

Sketch of the machine
(Current directions are obtained by Fleming's R.H. Rule)

F due to STATOR winding
max. value = 200×0·1 = 20A

F due to ROTOR winding
Max. value = ½×6×2×2 = 12A

F diagram due to currents on both winding

Fig. 6.18 Appertaining to Question 1

NOTE: (i) The *F* wave of the rotor winding is shifted by $\pi/2$ radians with respect to the *F* wave of the coil winding.

(ii) The flux density curve will not have the same shape as the m.m.f. curve because the air-gap is not uniform.

QUESTION 2. A machine consists of a stator carrying a three phase winding equivalent to two salient poles and a cylindrical rotor with a two-pole distributed coil winding as shown in the sketch (Fig. 6.19). At an instant when the current in RR' phase is a maximum, the m.m.f. axis of the rotor winding is $\pi/3$ radians ahead of the central line through the stator's RR' winding.

The maximum value of the m.m.f. F_S due to the three phase winding is 100 A and the maximum value of the rotor's m.m.f. F_R = 600 A. Find the resultant m.m.f. F_A in the machine's air-gap, if the rotor revolves in an anticlockwise direction at 3000 rev/min and the frequency of the currents in the stator is 50 Hz.

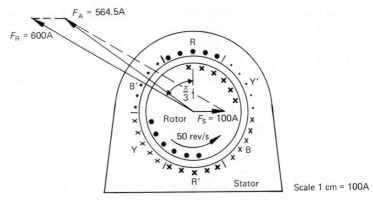

Fig. 6.19

ANSWER. The three phase winding produces an m.m.f. F_S which revolves at the same speed as the frequency of the currents flowing in it, i.e., at 50 revolutions per second.

The flux density B due to F_S has sinusoidal distribution and therefore F_S can be represented by a phasor revolving in space at 50 rev/s. The rotor m.m.f. F_R also produces a sinusoidal flux density in a uniform air-gap which however is stationary relative to the rotor's surface. But the rotor revolves at $\frac{3000}{60}$ rev/s = 50 rev/s, i.e., at the same speed as F_S. Hence phasors F_R and F_S can be added together because their relative position to each other is given and they revolve at the same speed, i.e., 50 rev/s. From the phasor diagram in Fig. 6.19

$$F_A = F_R + F_S = \underline{564 \cdot 5 \text{ A}}$$

and produces in the machine's air-gap the resultant magnetic flux density which is sinusoidal and revolves at 50 rev/s.

6.15 No-load e.m.f. of rotation in a distributed winding

In paragraph 3.11 of Chapter 3 a *single coil* sinewave machine was described. A practical machine however employs many such coils connected in series to form a distributed winding.

The effects which this produces are not evaluated simply by multiplying the behaviour of one coil by their increased number. Some new aspects also appear which are considered below.

The simple machine of Chapter 3 is used again here, but the number of its coils is trebled. The machine is shown in Fig. 6.20(a) with conductors A and C' permanently connected to the sliprings. The angle of spread between the conductors is made equal to $\pi/3$ radians. In Fig. 6.20(b) the machine is redrawn in the developed form to indicate clearly the interconnection between the conductors. In Fig. 6.20(d) the waveforms of the e.m.f.'s induced in the three coils are plotted against time. The e.m.f.'s are all sinewaves because the flux density curve which the coils 'cut' is sinusoidal and the rotor's speed is constant. The e.m.f. in coil AA' leads by $\pi/3$ rad. that in coil BB', whereas that in coil CC' lags behind it by $\pi/3$ rad.

The coils being connected in series produce a total winding e.m.f. which is the sum of their three e.m.f. waves. The result is a fourth sinewave. At the instant shown in Fig. 6.20(a) and (b) the coil BB' passes directly under the centres of the main poles and therefore cuts the maximum flux. The instantaneous value of the e.m.f. induced in the coil BB' is thus a maximum.

That instant of time is indicated by a vertical line XX' in Fig. 6.20(d) and the phasor diagram in Fig. 6.20(e) is drawn for the same time. Since $E_{BB'}$ and E_R are maximum at XX', their phasors are drawn vertically upwards. The flux Φ is added to the diagram in Fig. 6.20(e) and is drawn horizontally to correspond with the machine's cross-section in Fig. 6.20(a). It is seen that Φ lags E_R by $\pi/2$ rad. and therefore corresponds to the phasor diagram for a pure inductive coil (see Chapter 2).

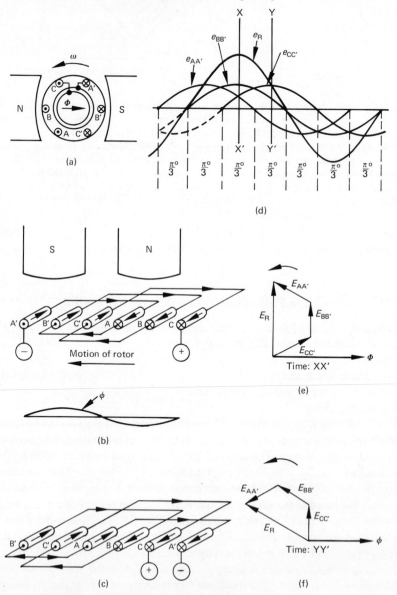

Fig. 6.20   Appertaining to 3-coil sinewave machine

When the rotor moves from right to left through $\pi/6$ rad., the e.m.f. in coil CC$'$ is now a maximum (Fig. 6.20(c)), and the instant of time when this occurs is shown by the line YY$'$ in Fig. 6.20(d)). It is clear that the e.m.f. of coil AA$'$ now opposes the e.m.f.'s in coils BB$'$ and CC$'$. Hence the instantaneous value of the resultant winding e.m.f. is less than its maximum. The above is seen from arrows in Fig. 6.20(c) as well as from the graph of the sinewaves, where the instantaneous e.m.f. of coil AA$'$ has a negative value.

The phasor diagram shown in Fig. 6.20(f) has now rotated through $\pi/3$ rad. in an anticlockwise direction, and the phasor $E_{CC'}$ is vertical indicating that the instantaneous value of the e.m.f. in coil CC$'$ is now a maximum. The flux Φ however remains horizontal because the main poles do not revolve. The main points to be noted are:

(i) The instantaneous values of the e.m.f.'s in all three coils are never equal i.e., at instant XX$'$ the coils CC$'$ and AA$'$ have lower e.m.f.'s induced in them than the coil BB$'$.

(ii) The maximum of the total winding e.m.f. is *less* than three times the value of each separate coil e.m.f.

(iii) The maximum of the total winding e.m.f. (resultant) occurs when the whole winding between the sliprings lies under the two poles.

(iv) The flux phasor Φ remains stationary and lags $\pi/2$ rad. behind the resultant winding e.m.f. E_R only at an instant when E_R is a maximum.

If the sliprings are replaced by two brushes resting on the commutator and the coil junctions are connected to the commutator segments, then the e.m.f. between the brushes will always be the same and equal to the maximum of the resultant sinewave, although the rotor revolves continuously.

The brushes ensure that the phasor diagram is always as in Fig. 6.20(e) and that the maximum instantaneous e.m.f. of the commutator winding always appears between them in the form of direct e.m.f.

In a phase winding the phasor diagram rotates anticlockwise and the winding e.m.f. between the sliprings is alternating.

6.16   Expression for e.m.f. of rotation in a distributed winding

In the alternating current machines described in this book the magnetic flux pattern in the air-gap is assumed to be sinusoidal as seen in paragraph 6.14 where the flux density waves due to distributed coil and phase windings were deduced. Furthermore, the resultant flux due to two or more windings of this type may also be regarded as sinusoidal.

This assumption is acceptable because a.c. machines are designed either to generate or to work from sinusoidal supply systems.

The expression for rotational e.m.f. induced in distributed windings is therefore deduced on the basis of sinusoidal flux distribution in an air-gap. The equation given in Chapter 3 for the e.m.f. of a single-turn winding is used as a starting point, i.e.,

$$E_r/\text{turn} = 4{\cdot}44\ k_p \Phi f \ (\text{V}) \tag{3.20}$$

If the winding distributed in slots on the surface of a rotor or stator has N turns connected in series, then it is known (paragraph 6.15) that the resultant e.m.f. is *not* N times larger than that due to one turn. In fact its value is somewhat smaller.

The equation (3.20) is therefore multiplied by N turns *and* a factor k_s which is smaller than unity. Thus the equation for e.m.f. of rotation of a distributed winding becomes

$$E_r/\text{winding} = 4{\cdot}44\, k_s k_p \Phi_A fN \text{ (V)} \tag{6.6}$$

k_s is termed the spread or distribution factor and is always less than unity, except when all the turns of the winding are placed in two slots a pole pitch apart Φ_A is the air-gap flux under no-load or on-load conditions.

The distribution factor k_s is derived by considering a section of the winding mounted within one pole pitch as follows:

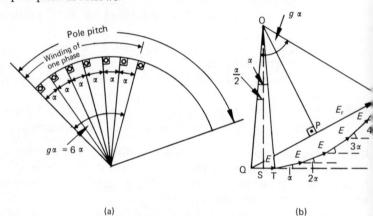

(a)   (b)

Fig. 6.21   Appertaining to spread factor

Let the winding be distributed in 'g' slots, α electrical degrees apart, over a fraction of a pole pitch (In Fig. 6.21(a) $g = 6$).

Therefore the angle subtended by the phase winding at the machine's centre = $g\alpha$ electrical degrees.

Let the r.m.s. value of the e.m.f. induced in the conductors lying in each slot be E volts.

Hence all the 'slot' e.m.f.s will be out of phase with each other by an angle $\alpha°$.

Since all the conductors are connected in series, then their resultant e.m.f. is equal to the phasor sum of all the e.m.f.s per slot.

The string phasor diagram is drawn in Fig. 6.21(b) where point 'O' is the intersection of perpendicular lines erected at mid-points of the phasors.

The spread factor k_s is the ratio of the resultant e.m.f. E_r to e.m.f. per slot multiplied by the number of slots.

hence
$$k_s = \frac{E_r}{gE} = \frac{2PQ}{g \times 2QS} = \frac{2 \times OQ \times \sin g\alpha/2}{g \times 2 \times OQ \sin \alpha/2}$$

i.e.,

$$k_s = \frac{\sin g\alpha/2}{g \times \sin \alpha/2} \tag{6.7}$$

QUESTION 3. A machine has three separate phase windings distributed in 36 slots around the stator's circumference.

Each winding is made up of fully pitched coils formed from 40 conductors accommodated in each slot.

A four pole rotor is driven at 25 rev/s and the resultant air-gap flux is sinusoidally distributed. Total flux per pole is 0·2 Wb.

Calculate (i) the spread factor

and (ii) the e.m.f. generated in each phase winding.

ANSWER. Number of slots per pole pitch $= \dfrac{36}{4} = 9$

Since there are three separate windings, each occupies only $\frac{1}{3}$ of the slots, i.e., $g = 3$ slots/pole.

The angle subtended by the three slots is thus $\dfrac{180}{3} = 60°$ and $\alpha = \dfrac{60°}{3} = 20°$

Hence spread factor $k_s = \dfrac{\sin(3 \times 20°)/2}{3 \times \sin 20°/2} = \dfrac{\sin 30°}{3 \times \sin 10°} = \underline{0·979}$

The coils being fully pitched, the pitch factor $k_p = 1$

Total number of conductors $= 36 \times 40 = 1440$

Total number of turns $= \dfrac{1440}{2} = 720$

$\therefore$ Number of turns per winding $= \dfrac{720}{3} = \underline{240}$

Rotor revolves at 25 rev/s, hence each conductor is swept by two pairs of poles in one revolution. Therefore its e.m.f. undergoes two complete cycles in one revolution. Therefore its e.m.f. undergoes two complete cycles in one revolution and the frequency $f = 25 \times 2 = 50$ Hz. Therefore e.m.f. per winding,

$$E_r = 4·44 \times 0·979 \times 1 \times 0·2 \times 50 \times 240$$
$$= 10\,440 \text{ V. or } \underline{10·44 \text{ kV.}}$$

SUMMARY

1. The pattern of magnetic flux in the air-gap determines the group to which a given machine belongs.

2. Windings are classified as

      (i) concentrated

and (ii) distributed.

3. Concentrated windings are invariably coil type.

4. Distributed windings may be

      (i) coil type

      (ii) cage type

      (iii) phase type

or (iv) commutator type.

5. The function of a slipring and brushgear is to transmit an electric current from a stationary to a rotating electric circuit.

6. The function of the commutator is to provide a connection between a stationary brush and the rotating commutator winding *and* to convert an alternating e.m.f. generated in the rotor conductors to a unidirectional e.m.f. (d.c.) between the brushes.

7.

$$\text{Flux density } B = \frac{\Phi}{A} \text{ (T)} \tag{6.1}$$

$$\text{Flux} \qquad \Phi = \frac{F}{S} \text{ (Wb)} \tag{6.2}$$

$$\text{Reluctance } S \quad = \frac{l}{\mu A} \text{ (A/Wb)} \tag{6.3}$$

$$\text{Flux density } B = \mu_r \mu_o \frac{F}{l} \text{ (T)} \tag{6.4}$$

$$\text{Flux density } B = \mu_o \frac{F}{l_g} \text{ (T) for the air-gap.} \tag{6.5}$$

where
$F = NI$ and is a magnetomotive force in A

l = the length of the magnetic paths in meters

l_g = the radial length of the air-gap in meters

A = the cross-sectional area in square metres

μ = the absolute permeability of the magnetic material

μ_r = the relative permeability

whilst
$\mu_o = 4\pi \times 10^{-7}$ H/m and is called the permeability of free space.

8. The m.m.f. distribution diagrams enable the flux pattern due to various types of windings to be determined. This in turn allows calculations of e.m.f. of rotation and the torque to be made.

9. E_r in a distributed winding for sinusoidal field pattern in the air-gap is given by

$$E_r = 4 \cdot 44 \, k_p k_s \, \Phi_A f N \text{ (V)} \tag{6.6}$$

where
k_p = pitch factor

Φ_A = flux per pole in Wb

f = frequency in Hz

N = the number of turns in series in a distributed winding

$k_s = \dfrac{\sin g\alpha/2}{g \sin \alpha/2}$ and is called the spread factor $\tag{6.7}$

g is the number of slots/pole pitch in which the winding is accommodated and α is the angle in *electrical* degrees between the adjacent slots.

EXERCISES

1. Distinguish between concentrated and distributed type of winding and give an example of each.

2. Describe with sketches each of the following windings:
   (i) concentrated coil winding,
   (ii) distributed coil winding,
   (iii) distributed cage winding,
   (iv) distributed phase winding,
   (v) distributed commutator winding.

3. Explain why each turn of a distributed winding is made up of a minimum of two conductors, which must be spaced approximately one pole pitch apart.

4. Describe a slipring and its brush gear. What is the purpose of this device?

5. Describe the construction of a commutator. What is its function?

6. Draw a developed diagram of a commutator winding given the following information:
STATOR: four salient poles; ROTOR: 12 slots;   two conductors per slot in two layers;   conductors form single turn coils connected in lap to 12 commutator segments.
   On your diagram show the position of the brushes and justify their placing.

7. Draw a developed diagram of a wave connected commutator winding using the information in question 6.
Note: Conductors in one slot and one commutator segment are not needed in this case.

8. Plot an m.m.f. diagram for the winding in question 6, assuming that each salient pole produces an m.m.f. of 26 A and each conductor carries 1·2 A.

9. A uniformly distributed two phase winding is placed on a stator. Each of its sections occupies 90° degrees of the stator's circumference and all the turns are fully pitched. When the two windings are fed from a two-phase balanced supply, draw m.m.f. diagrams at the following instants:

   (i) when the current in phase A is maximum positive,
   (ii) when the current in both phases is 0·707 of its maximum value, (the instant is equivalent to time interval of 45° after the instant in (i)),
   (iii) when the current in phase B is a maximum positive, (the instant is equivalent to time interval of 90° after the instant (i)).

Assume a straight line m.m.f. transition between the conductors in the slots.

10. Deduce an expression for the spread factor k_s.
Calculate the value of the spread factor for a single phase distributed winding with 10 slots per pole.
Answer: 0·639.

11. The distribution of flux density in a generator is sinusoidal, being maximum in the centre under the poles.
The machine has four slots per pole.
Derive curves representing the waveforms of the generated e.m.f. and its phasor diagram, when the winding is

   (i) concentrated in one slot per pole,
   (ii) distributed in two adjacent slots per pole,
   (iii) distributed in the four slots per pole.

12. A 3-phase synchronous machine is connected to a 50-Hz supply and has the following particulars,

Number of poles	4
Number of slots per pole/phase	6
Number of conductors per slot	8

The conductors of each phase are all connected in series to form a fully pitched winding. Flux is sinusoidally distributed in an air-gap and equal to 90 mWb per pole. Determine the r.m.s. value of the phase e.m.f. generated by the machine. Find also the line e.m.f. if the windings are connected in star.
Answer: 1828 V,   3162 V.

7 Direct current machines

This Chapter describes the theory and operation of different types of direct current machines. In each case, basic electromechanical power convertion equation, namely $T\omega = e_r i$ is emphasised. In order to stress the reversibility of the machine, that is, its ability to convert electrical power to mechanical power and vice versa, both generator and motor operation is given equal prominence.

Nevertheless, at the present time, industrial application of a d.c. machine as a generator is minimal as compared with its use as a d.c. motor for variable speed drives.

7.1 Definition of a d.c. machine

The discussion of the common features of all rotating machines showed that machines differ basically in the arrangement of windings and in the nature of the current flowing in them. These differences may be used for purposes of definition as follows:

'A d.c. machine has one or more concentrated coil windings on its STATOR, which carry direct current, and a commutator winding on its ROTOR, with direct current flowing either into or out of it'.

In the subsequent discussion of d.c. machines the rotor is often referred to as the armature.

7.2 Construction of a d.c. machine

Figure 7.1 shows a typical d.c. machine incorporating all the constructional details. The legend gives the name of each part.

7.3 Simple explanation of operation of d.c. machines

When a d.c. machine is operated either as a motor or as a generator, the coil winding is always supplied with direct current. This current produces an m.m.f., which in turn establishes a constant radial flux in the air-gap.

For a machine to operate as a generator, mechanical energy is supplied to the shaft of the rotor by a prime-mover (petrol engine, diesel engine or even human power may be used). The prime-mover turns the rotor and its commutator winding continuously. The conductors placed in slots 'cut' the magnetic field and an e.m.f. of rotation is induced in each. Since the conductors are connected in series in each parallel path,

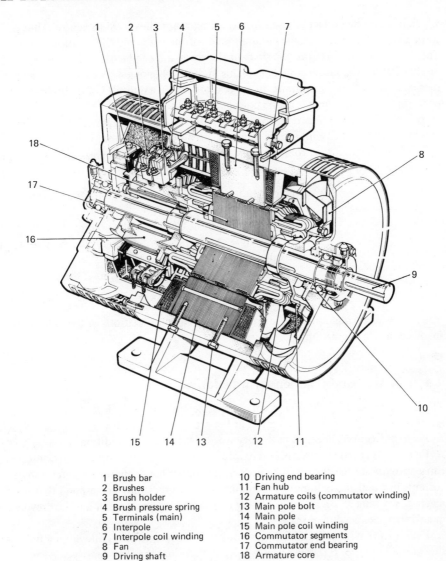

1 Brush bar	10 Driving end bearing
2 Brushes	11 Fan hub
3 Brush holder	12 Armature coils (commutator winding)
4 Brush pressure spring	13 Main pole bolt
5 Terminals (main)	14 Main pole
6 Interpole	15 Main pole coil winding
7 Interpole coil winding	16 Commutator segments
8 Fan	17 Commutator end bearing
9 Driving shaft	18 Armature core

Fig. 7.1   A typical d.c. machine with a section removed. (*Mawdsley's Ltd.*)

the sum of the e.m.f.s appears across the brushes, and will drive the current through an electrical load which is connected to the machine. In Chapter 3 electrical phenomena occurring in a single turn coil are discussed in detail. The d.c. machines described here can be regarded as consisting of a large number of such coils connected together so that the total effect obtained is very much greater although the principles involved are the same.

For the machine to operate as a motor, electrical energy must be supplied to the rotor. The direct current is fed from an external source via brushes into the conductors of the rotor. As a result each conductor carrying current and placed in the magnetic field produced by the coil windings, experiences a force. This force acts on conductors which lie near the surface of the rotor at a common radius from its centre. Thus a torque is produced around the circumference of the rotor and the rotation results. This torque is transferred to the shaft of the rotor and can be utilised to drive a mechanical load. Chapter 3 again explains in detail phenomena occurring in a single turn coil under these conditions. In a d.c. motor the total result is a multiple of effects occurring in a large number of the coils connected together.

It is clear from the above that whichever mode the machine is operated at, the conductors of the commutator winding (i) carry a current, (ii) experience a force and (iii) have an e.m.f. of rotation induced in them. Thus the expression 3.14 derived in Chapter 3 can be applied to an air-gap region of a d.c. machine as follows:

Let E_r = average e.m.f. of rotation induced in the commutator
                              winding.
                       I_a = total armature current flowing in the commutator winding.
                       T = average torque at which electro-mechanical power conversion
                              takes place, and
                       ω = constant speed of rotation of the machine, then

$$E_r I_a = T\omega \tag{7.1}$$

The succeeding paragraphs are devoted to detailed considerations of the quantities involved in this expression.

7.4   M.M.F. and flux density waveforms

The large number of coils used in a d.c. machine necessitates a closer look into the simple explanation of operation given in the previous paragraph.

An m.m.f. and magnetic flux density waves illustrate the conditions obtaining.

Figure 7.2(a) shows a two pole d.c. machine in a developed form. Both windings carry currents so that the machine is operating under load conditions. Directions of currents in a coil and commutator windings are such that they correspond to a machine working as a generator rotating in a clockwise direction or a motor revolving in an anticlockwise direction.

Figure 7.2(b) shows an m.m.f. wave due to the concentrated coil winding only. As shown in Chapter 6, flux density $B = \mu_o NI/l_g$ therefore the flux density wave is obtained by dividing the m.m.f. curve by the length of an air-gap l_g and multiplying it by a constant μ_o. Since l_g is constant under the salient poles, the flux density wave is practically the same as the m.m.f. wave, except at the pole tips, where flux fringing makes the transition from maximum to zero values less sharp and results in rounding the sharp corners of the rectangles.

When the machine operates under no-load conditions, flux density distribution in the air-gap shown in Fig. 7.2(b) is just of this form. Figure 7.2(c) shows an m.m.f. wave due to the commutator winding. The wave is shifted $90°$ with respect to the salient pole m.m.f. The two curves are said to be in *quadrature*, and remain stationary, although the armature rotates continuously.

Again dividing the m.m.f. wave by the length of the air-gap results in the flux density curve. Under the main poles the wave follows the straight line of the m.m.f.

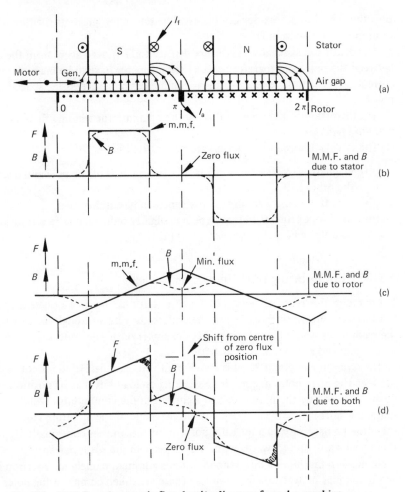

Fig. 7.2   M.M.F. and magnetic flux density diagrams for a d.c. machine

curve. Between the poles, however, the air-gap is very long and the flux density is reduced progressively until it reaches minimum value in the centre between the poles. Currents flowing in a commutator windings thus produce a magnetic flux in the interpolar space, where its effects are harmful.

In Fig. 7.2(d) both flux density waves produced by the currents in the two windings are combined to give a flux distribution in the machine's air-gap under loading conditions.

It is seen that the flux density is distorted, being increased over one half of the pole and decreased over the other half. The increase is nearly equal to the decrease and so the total magnetic flux per pole is almost the same, as when the machine is running unloaded.

The small reduction in total flux per pole, however, becomes more pronounced with increasing load on the machine (larger current in the commutator winding). The reduction is shown by a shaded portion in Fig. 7.2(d) and is due to saturation of the pole-tips where high flux density occurs. The saturation prevents the flux increasing in proportion with m.m.f.

The distortion also causes zero flux density to be moved away from the centre between the poles and therefore special arrangements have to be made to enable the current reversals to take place easily as conductors move across the space between the poles.

The distortion of the no-load flux by the commutator winding flux is referred to as *armature reaction.*

The effects of armature reaction are summarised below:

  (i)   magnetic flux density is increased over one half of the pole and decreased over the other half.
  (ii) zero flux density occurs off the centre between the poles.
  (iii) total flux produced by each pole is slightly reduced due to saturation at the pole tips where the flux density is high.

7.5   Commutating poles in d.c. machines

In Chapter 6 the 'rectifier' action of a commutator has been described in detail. The brushes were shown to short-circuit the coils as they pass through the centre between the main poles where the magnetic flux density is zero. The coils, therefore, have no e.m.f.s induced in them at that instant.

However, in paragraph 7.4 it is shown that zero flux density does not occur midway between the main poles in a machine working on load but that its position is shifted from the centre by the m.m.f. due to a current in the commutator winding. Furthermore, the shift is variable and depends on the value of the armature current. It would therefore be necessary to adjust the position of the brushes continually by rocking them backwards and forwards with every change in the rotor current.

In modern machines a different solution is adopted, namely an insertion of auxiliary poles between the main poles. These are called commutating poles, compoles or interpoles. Each carries a concentrated coil winding which is connected in series with the commutator winding so that the current which causes the distortion is used to neutralise it. The m.m.f. due to a compole is arranged to oppose that produced by the armature. Figures 7.3(a), (b), (c), and (d) illustrate the action of the compoles.

Figure 7.3(a) shows a portion of the machine illustrated in Fig. 7.2 with the compoles added.

In Fig. 7.3(b) an m.m.f. wave due to a commutator winding is re-drawn and in Fig. 7.3(c) the m.m.f. produced by the compole winding is shown. The ampere-turns of the compole are arranged to be equal to the ampere-turns of the commutator winding, so that in Fig. 7.3(d) the sum of the two equals zero in the region of the brush position. Thus the magnetic flux density in that region is zero and the null position is restored to the centre between the main poles.

The coils of the commutator winding when moving through this region do not 'cut' magnetic flux and hence have no e.m.f. induced in them, just as in a machine working on no load.

In practice the ampere-turns of the compoles are arranged to over-compensate the

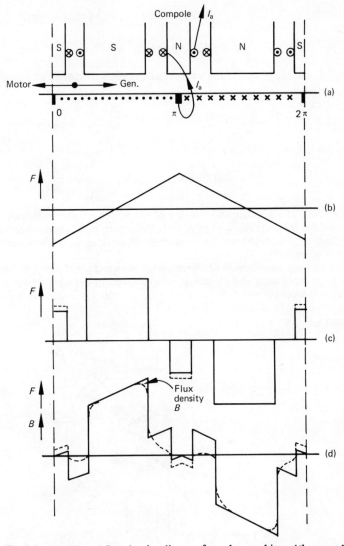

Fig. 7.3   M.M.F. and flux density diagram for a d.c. machine with compoles

armature ampere-turns (dotted line in Fig. 7.3(c) and (d)) to assist the change in current flow in the conductors undergoing commutation. The ampere-turns on each commutating pole are usually 1·2–1·3 times the number of armature ampere-turns/pole.

The armature ampere-turns between the main poles are due to current flowing in conductors lying in the space of one pole pitch, and are given by:

$$\text{Armature ampere-turns/pole} = \tfrac{1}{2} \times \frac{Z}{2p} \times \frac{I_a}{c} \tag{7.2}$$

where $\dfrac{Z}{2p}$ is the number of conductors per pole pitch.

Therefore $\quad\quad\quad\quad\quad \frac{1}{2} \times \frac{Z}{2p}$ is the number of turns per pole pitch

and $\quad\quad\quad\quad\quad\quad \frac{I_a}{c}$ is the current in each conductor.

To neutralise the m.m.f. due to these the interpole ampere-turn must be equal and opposite to them.

i.e., $\quad\quad\quad\quad$ Interpole ampere-turns $= \frac{Z}{4p} \times \frac{I_a}{c}$.

Since the interpole winding carries current I_a the number of turns on each is given by:

$$\text{Number of turns on each compole} = \frac{Z}{4pc} \quad\quad\quad (7.3)$$

QUESTION 1. Calculate the number of turns on each compole which would be required for a 4-pole lap-wound d.c. machine with 640 armature conductors, so that the effect of armature reaction is completely neutralised in the region where the brushes are placed.

Sketch the cross-section of the machine showing main poles, interpoles and commutator winding and electrical interconnection between them. Assume that the machine is working as a generator rotating in a clockwise direction.

ANSWER. No. of turns on each interpole $= \dfrac{640}{4 \times 2 \times 4} = \underline{20}$

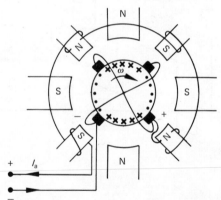

Fig. 7.4 Sketch of the machine showing polarity of the interpoles

7.6 E.M.F. of rotation and open circuit characteristic of a d.c. machine

Whether the machine is operating as a generator or as a motor, the commutator winding is rotating in the magnetic field which exists in the air-gap. The conductors of the winding 'cut' the magnetic flux and therefore the e.m.f. of rotation is induced in them. In the case of a generator, the e.m.f. of rotation is known as 'generated e.m.f.'

$$E_r = E_g.$$

In the case of a motor, the e.m.f. of rotation is known as 'back e.m.f.'

$$E_r = E_b.$$

The expression however is the same for both conditions of operation. To deduce it a d.c. machine operating on load is considered.

Let p = number of pairs of poles

                                Z = total number of conductors in a commutor winding

and c = number of parallel circuits through the winding between positive and negative brushes.

Therefore $\dfrac{Z}{c}$ = number of conductors in series in each parallel circuit.

where c = 2 for a wave winding

and c = $2p$ for a lap winding.

Figure 7.5 shows a portion of the machine embracing one pole pitch together with its flux density curve.

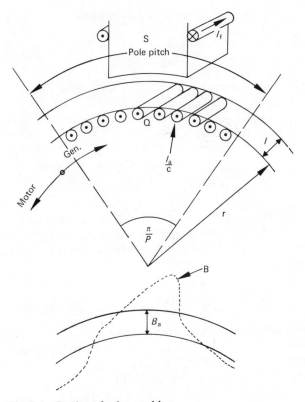

Fig. 7.5   Portion of a d.c. machine

If B_a is an average flux density over the pole pitch, then the average e.m.f. of rotation induced in one conductor Q is given by:

$$E_r \text{ per conductor} = B_a l \omega r \text{ (V)}$$

where l is an axial length of the armature in meters
         r is a radius of the armature in meters
and ω is the speed of rotation in radians per second.

As the conductor Q moves under each pole in turn, the same average value of e.m.f. of rotation is generated in it, and so the e.m.f. is maintained as long as the machine is rotating.

As the number of conductors between the brushes remains constant and equal to Z/c in each parallel circuit, the total e.m.f. of rotation is:

$$E_r = B_a l \omega r \frac{Z}{c} \text{ (V)} \qquad (7.4)$$

It is usual to express E_r in terms of the total flux per pole rather than the average flux density B_a, thus:

flux per pole $\Phi =$ Average flux density x area of the pole

$$= B_a \times l \times \frac{\pi}{p} \times r \text{ (Wb)}$$

∴
$$B_a = \frac{p\Phi}{\pi l r}$$

Substituting for B_a an equation (7.4)

$$E_r = \frac{p\Phi}{\pi l r} \times l \omega r \times \frac{Z}{c} = \frac{p\Phi\omega Z}{\pi c} \text{ (V)}$$

hence:

$$E_r = \frac{pZ}{\pi c} \omega \Phi \text{ (V)} \qquad (7.5)$$

It is seen from equation (7.5) that the e.m.f. of rotation is directly proportional to the flux per pole Φ and the speed of rotation ω the factor $pZ/\pi c$ being a constant for a given machine. If n = speed of rotation in revolutions per second then $\omega = 2\pi n$ and equation (7.5) becomes

$$E_r = 2p\Phi \times \frac{Z}{c} \times n \text{ (V)} \qquad (7.6)$$

Note that $2p\Phi$ = total magnetic flux of the machine.

When the machine is unloaded the flux per pole is produced by the current flowing in a concentrated coil winding around the salient poles as shown in Fig. 7.2(b).

This current is given the name of field or exciting current and is designated by I_f.

For a user of d.c. machines the relationship between e.m.f. of rotation E_r and field current I_f is of greater importance than the equation (7.5). Since it cannot be expressed by a formula, it is shown graphically in Fig. 7.6.

The curves are obtained experimentally by running a d.c. machine as a generator at a series of constant speeds. The machine is unloaded, the field current is varied from zero to a maximum, the generated e.m.f. E_r is measured and the results plotted.

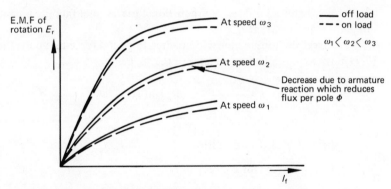

Fig. 7.6 Open Circuit characteristics of a d.c. machine

The curves thus obtained are termed Open Circuit characteristics of the machine and apply equally to its operation as a motor or a generator. When the machine is loaded the flux per pole is modified by armature reaction as shown in Fig. 7.2(d). The distortion results in a slight decrease in total flux per pole due to saturation at the pole tips (shaded portions) and therefore the e.m.f. of rotation is slightly reduced.

Hence the curves in Fig. 7.6 are modified to those shown by the dotted lines. It must be noted that the heavier the load on the machine (larger current in the commutator winding) the bigger the reduction in flux per pole and the greater the separation between the O.C. characteristic and its 'On Load' curve.

7.7   Torque of a d.c. machine

When a d.c. machine is loaded either as a motor or as a generator, the conductors of the commutator winding carry a current.

Since they all lie in the magnetic field of the air-gap, each one experiences a force and the torque is exerted around the circumference of the rotor.

When the machine is operating as a generator at constant speed this torque is equal and opposite to that provided by the prime-mover; when the machine is used as a motor, the torque is imparted to the mechanical load. If the speed in the latter case is constant the mechanical load torque is equal to the electrical torque developed.

The expression for the torque is the same for both modes of operation and is deduced from equation (7.1), i.e.,

$$T = \frac{E_r I_a}{\omega}$$

Substituting equation (7.5) for E_r,

$$T = \frac{E_r I_a}{\omega} = \frac{pZ}{\pi c} \times \omega \Phi \times \frac{I_a}{\omega}$$

or

$$T = \frac{pZ}{\pi c} \times \Phi \times I_a \ (\text{Nm}) \tag{7.7}$$

The torque therefore is directly proportional to the total flux per pole Φ and the armature (rotor) current I_a.

To convert the torque in newton-metres to pound force-feet units the expression (7.7) must be multiplied by a factor of 0·738, i.e.

$$T = \frac{0 \cdot 738\, pZ}{\pi c} \times \Phi I_a \text{ (lbf ft)} \tag{7.8}$$

7.8   Speed of a d.c. machine

Solving equation (7.6) for n gives:

$$n = \frac{c}{2pZ} \times \frac{E_r}{\Phi} \text{ (rev/s)} \tag{7.9}$$

Therefore the speed of rotation of a d.c. machine is proportional to the e.m.f. of rotation E_r and inversely proportional to the flux per pole Φ.

Since the expression for e.m.f. of rotation applies equally to motors and generators, equation (7.9) gives the speed for both modes of operation.

In the case of a d.c. motor the expression enables calculations of the speed at which the machine will run given the knowledge of E_r and Φ. In the case of a d.c. generator, the speed of a prime-mover can be estimated so that a required e.m.f. of rotation is generated at a certain value of flux Φ.

QUESTION 2. A four pole lap-wound d.c. machine has 80 slots on the rotor, with 8 conductors per slot. The machine is driven at 20 rev/s and the useful flux per pole is 30 mWb.

(i) Calculate the value of the generated e.m.f., and
(ii) the prime-movers torque if the armature current is 50 A.

ANSWER. Total number of conductors Z = 80 x 8 = 640;
$$p = 2;$$
$$c = 4;$$
$$\Phi = 0 \cdot 030 \text{ mWb};$$
$$\text{and } \omega = 2\pi \times 20 = 40\,\pi \text{ rad/s.}$$

Substituting in equation (7.5)

$$E_g = E_r = \frac{2 \times 640}{\pi \times 4} \times 40\pi \times 0 \cdot 03 = \underline{384 \text{ V.}}$$

Using expression (7.7)

$$T = \frac{2 \times 640}{\pi \times 4} \times 0 \cdot 030 \times 50 = \underline{152 \cdot 5 \text{ Nm}}$$

or $T = 0 \cdot 738 \times 152 \cdot 5 = \underline{113 \text{ lbf ft.}}$

Alternatively since E_r is already calculated, expression (7.1) can be used thus:

$$T = \frac{E_r I_a}{\omega} = \frac{384 \times 50}{40\pi} = \underline{152 \cdot 5 \text{ Nm}} \text{ as before.}$$

7.9   Graphical symbols for d.c. machines

Rather than draw various parts of electrical machines in detail the symbols shown in Fig. 7.7 are used.

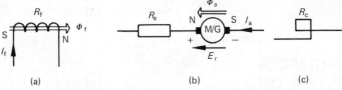

Fig. 7.7   Symbols for d.c. machines

Symbol (a) represents the main concentrated coil winding of resistance R_f.

The current I_f flowing through it produces a magnetic flux in the direction in which the current progresses through the winding.

Symbol (b) represents the rotor or armature and indicates:

(1) brushes,

(2) resistance R_a of the commutator winding including the resistance of brushes, commutator segments and brush connections,

and   (3) ideal source of e.m.f. of rotation E_r, generated in the commutator winding.

Again the armature current I_a produces the flux in the direction of the current flow through the winding in accordance with convention used in this book.

Symbol (c) represents the concentrated winding on the compoles, the resistance of which is R_c.

7.10   Interconnection of two windings in d.c. machines

The *simplest* d.c. machine consists of a minimum of two windings; one a concentrated coil and the other a commutator. The way they are connected to each other and to the supplies produces different types of d.c. machines which satisfy a variety of operating conditions.

Based on the methods of interconnection the following types are distinguished:

(i) separately excited d.c. machines

and   (ii) self exciting d.c. machines, with

(a) SHUNT connection

(b) SERIES connection.

7.11   Separately excited d.c. machines

In this machine the two windings form separate circuits.

It has been mentioned before that the coil winding does not have an e.m.f. of rotation induced in it but that its function is to produce a main magnetic field, i.e., to 'excite' the machine. The name 'separate excitation' is given because the coil or an exciting winding is fed from a d.c. supply which is not connected to the commutator winding. Figures 7.8(a) and 7.8(b) show diagrams of connections for a d.c. generator and a d.c. motor respectively.

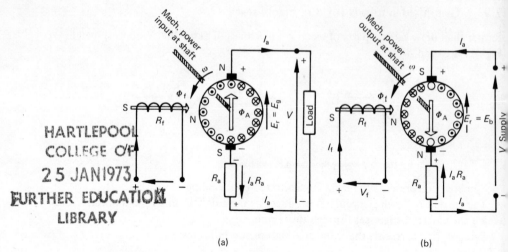

Fig. 7.8   (a) Generator (b) Motor. Separately excited d.c. machine

Considering the generator circuit marked according to the convention, the two voltage equations, one for each winding, are obtained by an application of Kirchhoff's voltage law. For the exciting winding circuit:

$$V_f = I_f R_f \tag{7.10}$$

and for the commutator winding

$$V = E_r - I_a R_a \tag{7.11}$$

Multiplying equation (7.11) by I_a:

$$VI_a = E_r I_a - I_a^2 R_a \text{ is obtained.}$$

But $E_r I_a = T\omega$, (equation (7.1))

therefore

$$VI_a = T\omega - I_a^2 R_a \tag{7.12}$$

where VI_a = electrical power output in W.

$T\omega$ = mechanical power input in W,

which is converted into electrical power, and

$I_a^2 R_a$ = electrical power lost in the resistance of the commutator winding in W.

Thus the mechanical prime-mover must supply more power to the machine than is obtained from it by an amount equal to $I_a^2 R_a$ loss.

The direction of rotation ω is clearly seen from the circuit, where the mechanical prime-mover must overcome the magnetic forces of attraction produced by the currents in the two windings.

The equation (7.10) applies to the generator under 'steady state' conditions; that is when it operates at a constant speed and a constant load. It does not apply during the time of load or speed changing.

Equations for the d.c. motor are obtained in the manner similar to that for the generator and are as follows:

For the exciting coil circuit:

$$V_f = I_f R_f$$

which is the same as equation (7.10),

and for the commutator winding:

$$V = E_r + I_a R_a \tag{7.13}$$

Multiplying (7.13) by I_a

$$VI_a = E_r I_a + I_a^2 R_a \text{ is obtained.}$$

again $E_r I_a = T\omega$

hence

$$VI_a = T\omega + I_a^2 R_a \tag{7.14}$$

where VI_a is the electrical power supplied to the rotor of the motor from a second d.c. source.

$T\omega$ is the mechanical power produced from electrical power, and $I_a^2 R_a$ is the power lost in the resistance of the commutator winding. Electrical Power VI_a exceeds the mechanical power obtained by the $I_a^2 R_a$ loss.

The direction of rotation is obtained from the circuit diagram by remembering that it is *produced* by the magnetic forces due to the currents in the windings. Therefore ω is in the direction in which the stationary 'N' pole of the coil winding attracts the 'S' pole of the rotor, i.e., anticlockwise.

Again the equations (7.13) and (7.14) apply only to steady state operation. An inspection of equations (7.11) and (7.13) and equations (7.12) and (7.14) shows that they differ only in the sign in front of the third term, the sign being minus (−) in the case of the generator and plus (+) in the case of the motor. This sign change is due to the reversal of the direction of the current I_a. If I_a is considered positive (+) when it flows *into* the commutator winding through a positive terminal and negative (−) when it leaves through the positive terminal the equations can be combined as under:

$$V = E_r \pm I_a R_a \tag{7.15}$$

and

$$VI_a = T\omega \pm I_a^2 R_a \tag{7.16}$$

NOTE: Voltage equation for the generator:

$$V = E_r - I_a R_a, \text{ when multiplied by } (-I_a) \text{ gives}$$
$$-VI_a = - E_r I_a + I_a^2 R_a$$

therefore to reduce it to the form given in (7.12) it has to be multiplied by -1, i.e.,

$$VI_a = E_r I_a - I_a^2 R_a.$$

It is also to be noted that direction of rotation is the same for both the generator and the motor operation of the d.c. machine considered, and so is the direction of current flow I_f through the exciting winding. It can be said therefore that the difference in machine's operation as a generator or as a motor reduces to *the difference in direction of current flow I_a through the commutator winding.* Therefore in subsequent circuits of d.c. machines, positive current through the commutator winding will indicate motor operation and negative current, generator operation, so that each connection can be considered with greater economy.

7.12   Self-excited d.c. machines

In these machines the two windings can either be connected in parallel, giving a SHUNT excited machine, or in series giving a SERIES excited machine.

SHUNT CONNECTION
Figure 7.9 shows a diagram of a shunt excited d.c. machine.

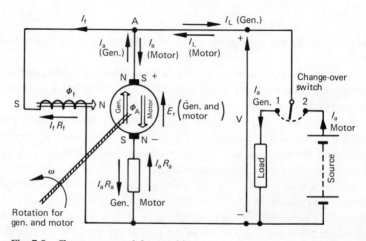

Fig. 7.9   Shunt connected d.c. machine

When the machine is driven by a mechanical prime-mover and the changeover switch is in position '1', the machine operates as a generator. The coil winding as well as the load are fed with electrical power produced in the commutator winding. When the changeover switch is in position '2' the electric power is supplied by a d.c. source to both windings of the machine and mechanical power is produced at the shaft.
The equations are as follows:

for the exciting winding

$$V = I_f R_f \tag{7.17}$$

and for the commutator winding

$$V = E_r \pm I_a R_a \tag{7.18}$$

The third equation relating the three currents, can be obtained by applying Kirchhoff's current law of the junction 'A':

for a motor case

$$I_f = I_L - I_a \tag{7.19}$$

and for a generator case

$$I_f = I_a - I_L \tag{7.20}$$

When the equation (7.18) is multiplied by I_a then

$$VI_a = E_r I_a \pm I_a^2 R_a \text{ is obtained.}$$

Substituting $T\omega$ for $E_r I_a$ the power equation results

$$VI_a = T\omega \pm I_a^2 R_a \tag{7.21}$$

Equation (7.21) is exactly the same as equation (7.16) for the separately excited machine.

QUESTION 3. A shunt excited machine is connected to 240-V. d.c. bus-bars. The machine's particulars are as follows:

> Number of poles = 6
> Flux per pole   = 50 mWb
> Resistance of the coil winding = 120 Ω
> Resistance of the commutator
>   winding incl. brushes etc.   = 0·1 Ω

The commutator winding is composed of 864 conductors connected in lap. Calculate:

(i) the speed and torque which the prime-mover must impart to the machine in order that it supplies 50 A to the d.c. bus-bars.
(ii) the speed and torque the machine would produce working as a d.c. motor taking 50 A from the bus-bars.

All losses other than resistive are to be neglected.

ANSWER. The current connections for a generator and a motor operation are shown in Fig. 7.10.

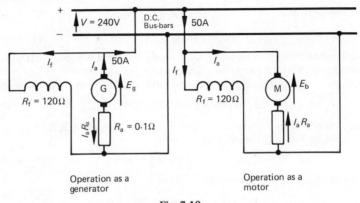

Fig. 7.10

Field current $I_f = \dfrac{240}{120} = 2$ A in both cases.

Considering operation as a generator first:

$$\text{Armature Current } I_a = I_L + I_f = 50 \text{ A} + 2 = 52 \text{ A}.$$

Hence using equation (7.18)

$$240 = E_g - 52 \times 0 \cdot 1$$

$\therefore$

$$E_g = 240 + 5 \cdot 2 = \underline{245 \cdot 2 \text{ V}}.$$

Equation (7.9) gives the speed

$$\omega = \frac{\pi c}{pZ} \times \frac{E_g}{\varPhi} = \frac{\pi \times 6}{3 \times 864} \times \frac{245 \cdot 2}{0 \cdot 05} = 35 \cdot 7 \text{ rad/s}.$$

or

$$n = \frac{35 \cdot 7}{2\pi} = 5 \cdot 68 \text{ rev/s}$$

and

$$\text{Torque} = \frac{E_g \times I_a}{\omega} = \frac{245 \cdot 2 \times 52}{35 \cdot 7} = 356 \text{ Nm}$$

or

$$0 \cdot 738 \times 356 = \underline{262 \text{ lbf ft}}$$

Considering operation as a motor:

$$\text{Armature current } I_a = I_L - I_f = 50 - 2 = 48 \text{ A}.$$

Hence from (7.18)

$$E_b = 240 - 48 \times 0 \cdot 1 = \underline{235 \cdot 2 \text{ V}}.$$

the speed of rotation

$$\omega = \frac{\pi \times 6 \times 235 \cdot 2}{3 \times 864 \times 0 \cdot 05} = \underline{34 \cdot 2 \text{ rad/s}}$$

or 5·45 rev/s

Alternatively it is seen that the speed is directly proportional to the e.m.f. of rotation, when the flux is unaltered.

Therefore $\omega = (\text{speed as a generator}) \times \dfrac{E_b}{E_g} = 35 \cdot 7 \times \dfrac{235 \cdot 2}{245 \cdot 2}$

$$= 34 \cdot 2 \text{ rad/s}.$$

and the Torque

$$= \frac{235 \cdot 2 \times 48}{34 \cdot 2} = \underline{328 \text{ Nm}}$$

or 242 lbf ft.

It is clear that the torque and speed, which the prime-mover must provide are greater than those obtained from the machine working as a motor. This is so, because in this example the power delivered or taken from the bus-bars is the same and equals:

$$VI_L = 240 \times 50 = 12\,000 \text{ W}.$$

The power lost in resistances R_f and R_a which is equal to $I_f^2 R_f + I_a^2 R_a$ must also be supplied by the mechanical prime-mover.

i.e., $2^2 \times 120 + 52^2 \times 0 \cdot 1 = 480 + 270 \cdot 4 = \underline{750 \cdot 4 \text{ W}}$

Thus the total power the prime-mover must supply equals

$$12\,000 + 750 \cdot 4 = \underline{12\,750 \cdot 4 \text{ W}}$$

In the case of the machine operating as a motor, the input is 12 000 W, and the losses are equal to

$$2^2 \times 120 + 48^2 \times 0\cdot1 = 480 + 230\cdot4 = \underline{710\cdot4 \text{ W.}}$$

Therefore the power converted to drive a mechanical load is only

$$12\ 000 - 710\cdot4 = \underline{11\ 289\cdot6 \text{ W.}}$$

SERIES CONNECTION

Figure 7.11 shows a diagram of a series excited d.c. machine. The machine operates as a generator when the change-over switch is in position '1' and the mechanical prime-mover drives the rotor in a clockwise direction. The machine works as a motor, taking a current from the supply when the change-over switch is in position '2'. The direction of rotation in this case is anti-clockwise since the current direction through the field coil winding as well as through the commutator winding is reversed. The current is given a symbol I in the figure. This is because the field winding, the armature winding and the load or the supply are all in series and the current I is the exciting current I_f, the armature current I_a and the load current I_L.

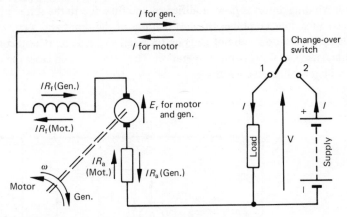

Fig. 7.11   Self excited – series connected d.c. machine

The equations deduced from the circuit are as follows:

$$I_f = I_a = I_L = I \tag{7.22}$$

and

$$V = E_r \pm I(R_a + R_f) \tag{7.23}$$

again multiplication of equation (7.23) by the current I results in the power equation

$$VI = T\omega \pm I^2(R_a + R_f) \tag{7.24}$$

The expression $I^2(R_a + R_f)$ is the power lost in the resistance of both windings connected in series.

VI is an electrical input or output of the machine, and $T\omega$ is a mechanical input or output according to the mode of operation.

7.13   D.C. machines with three windings (compound connection)

In previous paragraphs d.c. machines consisting of only two windings are considered.
The windings are

> (i) a concentrated coil (field) winding
and   (ii) a distributed commutator (armature) winding.

In some machines however a third winding is added to the salient poles, so that each
pole now carries two. The extra winding is termed a series field.

One field coil, the main one, is connected in parallel or shunt with the commutator
winding and the other in series with the commutator winding. Such a machine is said
to be a self-excited compound connected d.c. machine.

A series winding must always carry a large armature current I_a and therefore is
made of large cross-section conductors and has few turns only. The resistance of the
winding is very small so as not to reduce the e.m.f. of rotation by a significant amount.

A shunt winding on the other hand has many turns made of a small cross-section
conductor. Its resistance is high and the current flowing through it is small by
comparison with the machine's armature current I_a. The main magnetic flux is provided
by this winding, but it is now modified by the flux due to the series winding.

Such a machine combines the best features of the shunt and series excited types.

Figure 7.12 shows a circuit diagram for a self-excited compound-connected d.c.
machine, marked for generator operation. The reader should label the circuit for
motor operation as an exercise.

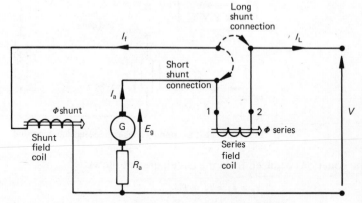

Fig. 7.12   Self excited – compound connected d.c. machine

It is seen in the diagram that there are two possible ways of connecting a 'shunt'
winding:

(i) across the commutator winding only, or
(ii) across both, the commutator winding and the series winding.

The first alternative is described as a 'short shunt' and the second as a 'long shunt'
connection.

Furthermore the fluxes due to two field coils in Fig. 7.12 are shown in the same
direction, i.e., the series winding flux strengthening the main flux produced by the

shunt winding. This arrangement is called 'cumulative'. If the connections to terminal '1' and '2' in Fig. 7.12 are reversed so that the current I_a flows through the series field coil from terminal '2' towards terminal '1', then the series flux is reversed and it now weakens the main field. This is called the 'differential' connection.

   In practice the first method is more commonly used of the two possibilities. The equation for compound machines can be deduced from the circuit diagram by an application of Kirchhoff's Laws in the same way as for shunt or series excited machines. It is, however, much easier to deal with each problem using numerical values then deducing general equations for each connection in terms of letters. This is shown in the following worked example

QUESTION 4. A 'short shunt' compound wound d.c. machine is connected to a 400 V d.c. supply and runs as a motor taking a current of 40 A. The machine's particulars are:

Armature resistance	0·21 Ω
Series field coil resistance	0·3 Ω
Shunt field coil resistance	194 Ω

Draw a circuit diagram, properly labelled and calculate:

   (i) shunt field current,
   (ii) armature current,
   (iii) back e.m.f. (e.m.f. of rotation).

If the speed of rotation is 20 rev/s., calculate the mechanical torque developed by the motor.

   All losses other than resistive losses are to be neglected.

ANSWER

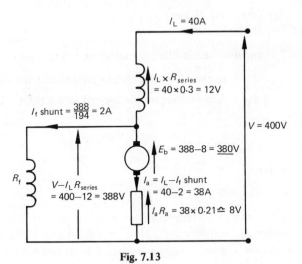

Fig. 7.13

The solution can be worked out in the diagram (Fig. 7.13), or as follows:

Potential difference on series field = 40 x 0·3 = 12 V
Potential difference across the shunt field = 400 − 12 = 388 V

$$\therefore \text{ shunt field current} = \frac{388}{194} = 2 \text{ A}$$

Current through armature $I_a = 40 - 2 = 38$ A
Potential across armature resistance $= 38 \times 0.21 = 8$ V
$\therefore$ E.M.F. of rotation = back e.m.f. $= 388 - 8 = \underline{380}$ V
Since $T\omega = E_r I_a$

$\therefore$ mechanical torque $= \dfrac{380 \times 38}{20 \times 2\pi} = \dfrac{361}{\pi} = 114.8$ Nm

7.14   Rotational losses in d.c. machines

The expression: $E_r I_a = T\omega$ shows that:

(1) in an electrical motor, electrical power of $E_r I_a$ (W) is converted to mechanical power equal to $T\omega$ (W)

and (2) in an electrical generator, the mechanical power of $T\omega$ (W) is converted to electrical power equal to $E_r I_a$ (W).

In practice it is found, however, that an electric motor produces *less* power for useful work than $T\omega$ (W), whereas a prime-mover driving an electrical generator has to provide *more* power than $T\omega$ (W) in order to obtain $E_r I_a$ (W) from it.

The difference in both cases is the *rotational loss C* (W), for as soon as rotation occurs a certain amount of power is necessary to overcome:

(i) friction in the bearings,
(ii) windage,
and (iii) losses in the magnetic core of the machine, which are subdivided into
(a) Eddy current loss,
and (b) Hysteresis loss.

The expression 7.1 may now be modified to include rotational loss C as follows:

Let T_M = an electric motor torque produced at the shaft
and T_P = prime-mover torque applied to the shaft of a generator.

Then for a d.c. motor

$$E_r I_a = T\omega = T_M \omega + C \qquad\qquad (7.25)$$

and for a d.c. generator

$$E_r I_a = T\omega = T_P \omega - C \qquad\qquad (7.26)$$

T in the above equations may be termed an 'internal Torque'. At this value of the torque the conversion of electro-mechanical power takes place. In d.c. motors it is sometimes referred to as a 'gross' torque whereas T_M is termed a 'nett' torque.

QUESTION 5. The shunt excited d.c. machine is run as an unloaded motor at its normal operating speed of 25 rev/s. It takes 3 A from a 240-V d.c. supply. The resistance of the field circuit is 240 Ω and the resistance of the armature circuit is 1 Ω. Calculate:

(1) back e.m.f.
and (2) rotational loss at this speed.

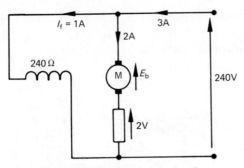

Fig. 7.14 Circuit diagram for Question 5

Field current $I_f = \dfrac{240}{240} = 1$ A

Armature current $I_a = 3 - 1 = 2$ A

Back e.m.f. $E_b = 240 - 2 \times 1 = \underline{238\ V}$

Using equation (7.25)

$$238 \times 2 = T_M W + C$$

but $T_M = 0$ because the machine is unloaded

∴ $C = 238 \times 2 = \underline{476\ W}$

Thus the input to the machine is expended entirely in resistive losses equal to $240 \times 1 + 2 \times 2 = 244$ W, and the rotational losses which amount to 476 W.

7.15   Friction, windage and core loss

BEARING FRICTION
This loss is due to rubbing of metal surfaces inside the bearings, although the effect is minimised by the use of lubricant. In modern machines bearing friction forms only a small portion of the losses and is roughly proportional to the speed of rotation 'ω'.

WINDAGE
This loss is due to power being used in setting up circulating currents in the air or other cooling gas in which the rotor with its fan revolves. The loss is proportional to the square of the speed, 'ω^2'.

CORE LOSSES
(a) *The 'Eddy current' loss* is due to currents induced in the ferromagnetic material of the machine's rotor as it rotates in the magnetic field of the stator. Every portion of the rotor's core behaves as a conductor 'cutting' a stationary magnetic flux and the e.m.f. induced in it obeys the equation $e_r = Bl\omega r$.

All the e.m.f.s drive the currents around the rotor and the power lost is equal to the sum of all 'i^2R's', where R is the resistance of each individual path through the core. Figure 7.15(a) shows what occurs in a solid rotor and Fig. 7.15(b) gives the section through the rotor, cut along the shaft, which is made up of a number of laminations. These are insulated from each other and about 0·4 mm in thickness, to break up the path of the eddy currents and thus reduce the power lost. For this reason all rotors of industrial machines are made up of laminations. The loss is proportional to the square of the speed, 'ω^2'.

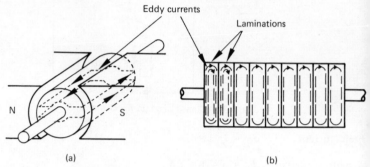

Fig. 7.15   Eddy currents in the rotor of a d.c. machine

(b) *The hysteresis loss.* which is the remaining part of the core loss is caused by the alternating magnetisation of the rotor. Groups of atoms form domains and each domain behaves as a very small permanent magnet. As such it aligns itself with the magnetic flux in which is it placed. As the rotor. Groups of atoms form domains and each it, but in addition it also rotates about its own axis. This 'movement' is clearly seen from Fig. 7.16 where six different positions of one domain during a revolution are shown.

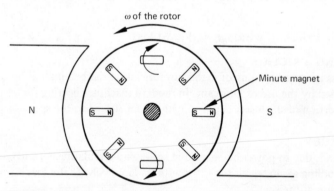

Fig. 7.16   Hysteresis Loss in the rotor

All the domains which go to make up the rotor's material undergo this cyclic magnetisation in each revolution. A certain amount of power is expended in the process which causes heating of the rotor. The power lost is called a 'hysteresis loss' and is proportional to the speed of rotation, 'ω'.

All these losses depend on the speed of rotation, and as long as the speed is kept constant they may be assumed to be the same irrespective of the load on the machine. In practice this is not quite true, for the value of the magnetic flux also affects them. The variation however, is allowed for by an extra amount of power lost referred to as a 'stray loss'. It is difficult to estimate this value however, and in this book the stray loss is neglected.

7.16   Efficiency of d.c. machines

Applying general equation (1.7) deduced in Chapter 1 to:

(1) a d.c. generator

$$\text{Efficiency } \eta = 1 - \frac{\text{all resistive losses} + C}{VI_L + \text{all resistive losses} + C} \text{ p.u.} \tag{7.27}$$

where C = rotational loss in W,

V = output voltage in V,

and I_L = load current in A,

so that VI_L is the generator output in W.

(2) a d.c. motor

$$\text{Efficiency } \eta = 1 - \frac{\text{all resistive losses} + C}{T_M \omega + \text{all resistive losses} + C} \text{ p.u.} \tag{7.28}$$

where T_M = output torque at the shaft and ω is the speed of rotation

so that $T_M \omega$ is the motor output in W.

The resistive losses include all or some of the following:

$I_a^2 R_a$ = the loss in the resistance of the commutator winding, brushes, leads, commutator segments and the contact resistance between the brushes and the commutator.

$I_a^2 R_c$ = the loss in the resistance of the interpole winding

$I_{f1}^2 R_{f1}$ = the loss in the resistance of the shunt field winding including field coil, instruments, rheostats, etc.

$I_{f2}^2 R_{f2}$ = the loss in the resistance of the series field winding.

QUESTION 6. The machine in Question 5 is rated at 5 kW. When the machine is working at full load and its nominal speed of 25 rev/s., calculate:

(1) The prime-mover's torque and the efficiency of the machine when working as a generator,

and (2) The output torque and the efficiency of the machine when working as a motor.

ANSWER. *Generator*

Full load current $= \dfrac{5000}{240} = 20.8$ A.

Shunt field current = 1 A.

Armature current   = 20·8 + 1 = 21·8 A.

Generated e.m.f.   = 240 + 21·8 x 1 = 261·8 V.

Using equation (7.26)

$$T_p\omega - 476 = 261{\cdot}8 \times 21{\cdot}8$$

$$\therefore T_p = \frac{261{\cdot}8 \times 21{\cdot}8 + 476}{25 \times 2\pi} = \frac{5710 + 476}{50\pi}$$

$$= \frac{6186}{50\pi} = 39{\cdot}4 \text{ Nm}$$

Output power   = 5000 W.

Losses:

Rotational Loss	= 476 W.
Shunt field Loss	$= VI_f = 240 \times 1 = 240$ W.
Armature loss	$= I_a^2 R_a = 21{\cdot}8^2 \times 1 = 475$ W.
Total Loss	$\overline{1191 \text{ W.}}$

Efficiency $= 1 - \dfrac{1191}{5000 + 1191} = 1 - 0{\cdot}193 = \underline{0{\cdot}807}$ p.u.

or 0·807 x 100 $= \underline{80{\cdot}7}$ per cent.

Motor

Output power   = 5000 W.

$$\therefore \text{Output Torque } T_M = \frac{5000}{25 \times 2\pi} = 31{\cdot}8 \text{ Nm.}$$

Using expression (7.25)

$$E_r I_a = 5000 + 476 = 5476 \text{ W.} \tag{1}$$

Since neither E_r nor I_a are known a second equation is necessary.

This is obtained from equation (7.18)

$$240 = E_r + I_a \times 1 \tag{2}$$

Substituting from (1) for I_a

$$240 = E_r + \frac{5476}{E_r} \times 1 \text{ or } E_r^2 - 240E_r + 5476 = 0.$$

Solving for E_r

$$E_r = \frac{240 \pm \sqrt{240^2 - 4 \times 5476}}{2} = 214{\cdot}5 \text{ V. or } 25{\cdot}5 \text{ V.}$$

Only the first answer is applicable in this case and therefore from equation (2)

$$I_a = \frac{240 - 214{\cdot}5}{1} = 25{\cdot}5 \text{ A.}$$

Losses:

Rotational C	= 476 W.
Shunt field	= 240 W.
Armature circuit $(25{\cdot}5)^2 \times 1$	= 640 W.
Total Loss	$\overline{1356 \text{ W.}}$

Efficiency $= 1 - \dfrac{1356}{5000 + 1356} = 1 - 0{\cdot}213 = \underline{0{\cdot}787}$ p.u.

or $\underline{78{\cdot}7}$ per cent.

7.17   Characteristics of d.c. machines

The knowledge of *how* the d.c. machines, discussed so far, behave when operating under various load conditions is very necessary to an engineer who uses them.

For a d.c. machine, *working as a generator*, must deliver an electrical power VI_L watts to the load at a voltage V volts and a current I_L amperes. How V varies with I_L is therefore an important question and can be answered experimentally by driving an appropriately excited machine at a constant speed, a constant value of a shunt field current and varying a loading resistance. The resulting values of terminal voltage V and the load current I_L, when plotted give a graph called 'a load or external characteristic' of a given generator. Figures 7.17(a) and (b) show typical load characteristics for various types of d.c. generators.

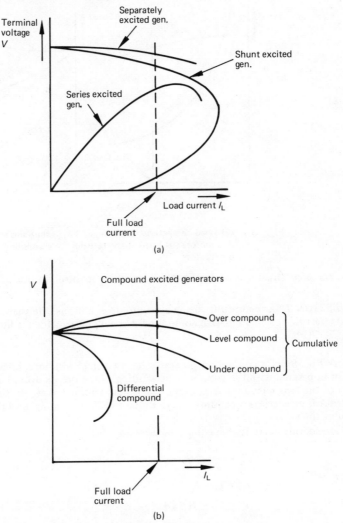

Fig. 7.17   Typical load characteristics of d.c. generators

Similarly the machine *used as an electric motor* must deliver mechanical power $T_M \omega$ watts to the load. Here it is important to know how the torque T_M (Nm) varies for different values of speed n (rev/s.) The graph of n plotted against T_M for various types of motors is obtained experimentally by loading the machine with a brake and keeping the supply voltage V and a shunt field current I_f constant.

The typical curves for various types of d.c. motors shown in Figs. 7.18(a) and (b), are called their 'Load or Speed/torque characteristics'. It must be noted that apart from experimental methods, the shapes of all the characteristics for generators and motors can be predicted from the equations deduced in paragraphs 7.11 and 7.12 and the knowledge of the effect of armature reaction described in paragraph 7.4.

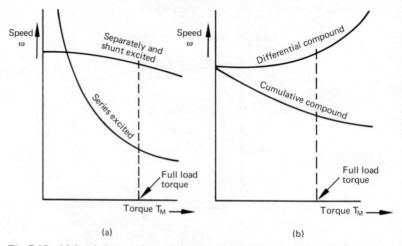

(a)   (b)

Fig. 7.18   (a) Load characteristics for separately, shunt and series excited
d.c. motors (b) Load characteristics for compound excited
d.c. motors

Examples which follow illustrate the reasoning involved in this process.

QUESTION 7. Explain fully, using appropriate equations, the shape of the load characteristic for a cumulative compound d.c. generator, which is illustrated in Fig. 7.17(b).

ANSWER. Since a compound generator has two field windings, one in series and the other in parallel with the commutator winding, therefore its overall characteristic is a sum of the two separate curves due to each field winding alone. In the figure 7.19, curve A is due to series winding, curve B due to shunt winding, and the curve C is the sum of the two.

Considering curve A, the equations used are:

$$E_g = \frac{pZ}{\pi c} \omega \Phi \tag{7.5}$$

$$I_f = I_a = I_L = I \tag{7.22}$$

and
$$V = E_g - I(R_a + R_f) \tag{7.23}$$

The value of E_g depends on Φ only when the speed ω is kept constant; Φ in turn is caused by I, which is the series field current as well as the load current. Thus the shape

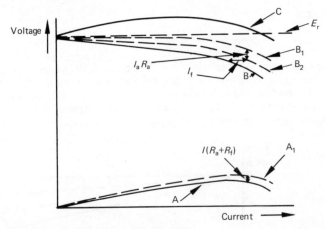

Voltage

I_aR_a

I_f

C

E_r

B_1

B_2

B

$I(R_a+R_f)$

A_1

A

Current →

Fig. 7.19 Appertaining to Question 7

of E_g is the same as for an open circuit characteristic, reduced by an armature reaction. This is curve A_1. From equation (7.23), V is less than E_g by $I(R_a + R_f)$, which therefore must be subtracted from curve A_1 to obtain the characteristic A. Now considering curve B, the appropriate equations are (7.5), as before,

and
$$I_f = I_a - I_L, \tag{7.20}$$
$$V = E_g - I_aR_a, \tag{7.18}$$
$$V = I_fR_f. \tag{7.17}$$

Again E_g is due to flux Φ, which is caused by I_f and slightly reduced by armature reaction. Thus curve B_1 represents E_g plotted against I_a. V lies below curve B_1 by a distance equal to I_aR_a. Finally to obtain V against I_L equation (7.20) shows that I_f must be subtracted horizontally from the curve B_2 to obtain the final characteristic B. It must also be noted that the curve B_1 dips slightly not only because of armature reaction, but also due to decrease of I_f, which is caused by decreasing V (equation (7.17)).

QUESTION 8. Explain the shape of the speed/torque characteristic of a shunt excited d.c. motor shown in Fig. 7.18(a).

ANSWER. The appropriate equations are:

$$I_L = I_a + I_f \tag{7.19}$$
$$V = E_b + I_aR_a \tag{7.18}$$
$$T = \frac{pZ}{\pi c} I_a\Phi = kI_a\Phi \tag{7.7}$$

and
$$E_b = \frac{pZ}{\pi c} \Phi\omega = k\omega \tag{7.5}$$

where
$$k = \frac{pZ}{\pi c}$$

Substituting (7.5) and (7.7) in (7.18)

$$V = k\omega + \frac{T}{k\Phi} R_a \text{ is obtained.}$$

Rearranging

$$\omega = \frac{1}{k\Phi} V - \frac{R_a}{k^2 \Phi^2} T.$$

In a shunt excited motor Φ is nearly constant because the supply voltage V is kept constant and therefore $I_f = V/R_f$ is also constant. Furthermore $1/k\Phi$ and $R_a/k^2\Phi^2$ are constants of which the second is very small. Therefore the speed drops gently with increase of torque T as shown below. The final curve of T_M is obtained by suptracting from curve T, rotational losses C.

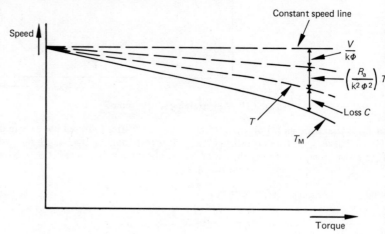

Fig. 7.20 Appertaining to Question 8

7.18   Control of d.c. machines – generators

E.m.f. of rotation E_r is given by an equation

$$E_r = \frac{pZ}{\pi c} \Phi\omega \text{ (V)} \tag{7.5}$$

which determines the eventual value of the terminal voltage V in all types of d.c. generators. Hence in order to control the voltage V the value of E_r must be capable of adjustment. Examination of the equation (7.5) shows that once the machine is built the only two quantities which can be varied are: the speed ω and the value of flux per pole Φ. The speed of rotation is determined by the mechanical prime-mover and therefore how it can be controlled depends on the particular mechanical drive used. The magnetic flux Φ on the other hand is due to the field current flowing through a concentrated coil winding, and therefore variation of the current I_f provides a simple means of controlling the value of e.m.f. of rotation.

A circuit diagram for a separately excited generator is shown in Fig. 7.21(a).

The rheostat R is connected in series with the field winding and the ammeter A_1 which indicates the value of I_f. The ammeter A_2 reads an armature current I_a and a voltmeter V the terminal voltage V which is applied to the load L.

The Fig. 7.21(b) shows a family of load characteristics, each obtained for a different setting of the rheostat R. When the rheostat is shorted the resistance of the

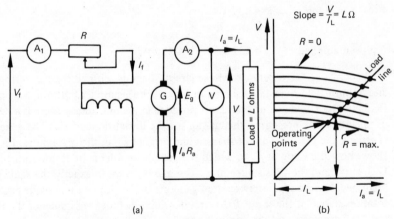

Fig. 7.21   (a) Circuit of separately excited d.c. generator (b) Load characteristics for different settings of R

field circuit is a minimum and equal to that of the winding alone. The field current I_f is consequently a maximum, the flux Φ also a maximum, and the value of E_r greatest. The other limit to E_r is set by the rheostat R having its full value in the circuit making I_f a minimum. Between these two limits there is an infinite number of values of E_r and therefore an infinite number of curves can be plotted of V against I_L.

To find an operating point for a d.c. generator from a given load characteristic and the value of the loading resistance, the load line is plotted on the same graph. The slope is made equal to the value of the resistance of the load, $L = V/I_L$ and the line is drawn from the origin of the graph. The line is shown in Fig. 7.21(b) and the intersection of it with the load characteristic gives an operating point.

QUESTION 9.   The load characteristic for a separately excited d.c. generator was obtained experimentally and is given below.

Terminal Voltage V (V)	240	238	236	233·5	230·5	228·5
Load current I_L (A)	0	5	10	15	20	25

Plot V against I_L and determine the value of voltage, current and the power the generator will supply to a load of 12·5 Ω.

ANSWER.

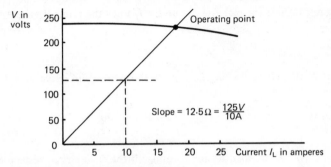

Fig. 7.22   Appertaining to Question 9

From the graph $V = 232$ V

 $I_L = 18 \cdot 6$ A

∴ power supplied $= 232 \times 18 \cdot 6 = 4320$ W

A *shunt excited generator*, whose circuit diagram with all the measuring instruments and a shunt field rheostat is shown in Fig. 7.23(a), supplies both the load and its own field circuit. The machine could not produce any e.m.f. unless there is a certain amount of magnetic flux present in the air-gap for the initial value of E_r to be generated. This is usually the case because of the residual magnetism in the main poles. Fig. 7.23(b) shows how the value of E_r to which the generator will excite is determined from an O.C. curve and the 'Resistance line'. The line is drawn in exactly the same way as the 'load line' for the separately excited generator. The slope of it is made equal to the total resistance of the shunt field circuit, which includes resistance of the field coil, rheostat, ammeter and the connections.

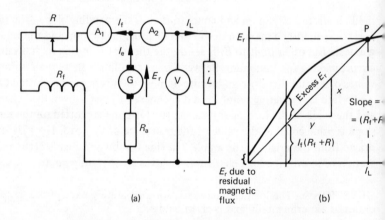

(a)   (b)

Fig. 7.23   (a) Circuit of a shunt excited generator (b) Dependence of E_r
on value of $(R_f + R)$

The point P at which the O.C. curve and the Resistance Line intersect, gives the position of equilibrium at which $E_r = I_f(R_f + R)$. Until this point is reached the O.C. curve lies *above* the resistance line and E_r exceeds the voltage drop $I_f(R_f + R)$. This excess is responsible for increase of current I_f until the point P, where the field current stops rising.

QUESTION 10. The following table gives an open-circuit curve of a shunt generator running at 20 rev/s.

E.M.F. of rotation E_r (V)	10	172	300	360	385	395
Field current I_f (A)	0	1	2	3	4	5

Determine (i) the no-load terminal voltage if the field circuit resistance is 125 Ω,
and   (ii) the critical resistance of the shunt field circuit.

ANSWER.

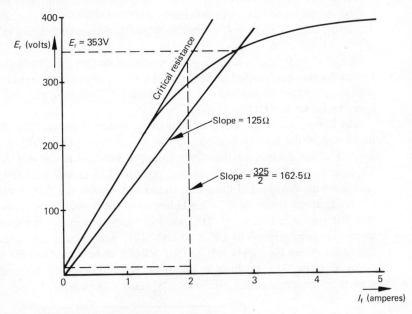

Fig. 7.24 Appertaining to Question 10

(i) From the graph E_r at R_f = 125 Ω is 353 V

(ii) Critical Resistance is that value of R_f which divides the build up of excitation of the generator from that at which the machine does not produce e.m.f. greater than a few volts due to residual magnetic flux only. It is given by the slope of the line tangential to the linear portion of the O.C. curve and its value from the graph is $\frac{325}{2}$ = 162·5 Ω.

The voltage of the compound wound d.c. generator is controlled by varying the exciting current through a shunt field winding only, much as for the shunt excited machine.

The effect of the series field winding is not normally altered at all after the machine has been built to its design specification.

7.19   Control of d.c. machines – motors

The *control* of a d.c. machine operating as a motor *involves starting it from rest* and the *ability to alter its speed of rotation* according to the requirements of the load.

STARTING

At an instant of switching the supply on, the e.m.f. of rotation E_r is zero and the equation (7.18) reduces to

$$V = I_a R_a$$

Assuming, as is quite common, that the supply voltage V is constant, the initial starting current is given by:

$$I_a \text{ at starting} = \frac{V}{R_a} \text{(A)} \qquad\qquad (7.29)$$

The resistance of armature circuit R_a is usually quite small and therefore the starting current would be very large; approximately 10 times its normal value. For instance if $V = 240$ V, $R_a = 1$ Ω, then I_a at starting is 240 A, whereas its normal value would be of the order of 20 A.

It is obvious from the foregoing that the initial current must be limited in most, except the smallest, machines. This is done by inserting an extra resistance in series with the commutator winding and then reducing it in stages to zero as the motor's speed builds up to its normal value.

The device, which consists essentially of the robust resistance 'S' to limit the initial rush of the current is called a starter and is shown in Fig. 7.25. It also serves as a switch which disconnects the supply from the motor when the arm 'A', pivoted at 'O', is in the 'OFF' position. At starting, the arm 'A' is moved manually from stud to stud, allowing the speed of the motor to build up until in the 'ON' position the supply is directly across the motor's both windings, and the resistance 'S' is completely cut out. The arm 'A' is held in the 'ON' position, against the pull of the spring 'C' by an electro-magnet energised by a No-volt-coil 'NVC' in series with the field winding of the motor. When the supply voltage fails 'NVC' is de-energised and the arm 'A' is returned to the 'OFF' position by the spring, thus ensuring that the motor is not started subsequently without resistance 'S' being in the circuit.

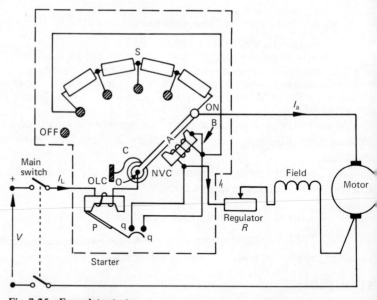

Fig. 7.25   Face plate starter

The connection marked 'B' enables the field current to by-pass resistance 'S' in an 'ON' position. Its path is from the terminal of the supply via arm 'A', the iron core of the electromagnet, the coil of NVC, to the field rheostat R.

The Over-load-coil 'OLC' protects the motor against excessive current I_a. At the predetermined value of I_a, the armature P is attracted by the core of the electro-magnet 'OLC' and closes contact qq, thus short-circuiting the NVC and releasing arm 'A' from the 'ON' position.

To stop the motor at will, the main switch is opened.

Thus the functions of the starter are:

- (i) to limit the starting current,
- (ii) to provide means of disconnecting the motor from the supply,
- (iii) to protect the motor from excess starting current in subsequent starts following the loss of supply voltage,
- (iv) to protect the motor from overload.

SPEED CONTROL

The speed of rotation of a d.c. machine is given by an equation (7.9) as

$$n = \frac{c}{2pZ} \times \frac{Er}{\Phi} \text{ (rev/s)}.$$

For all d.c. motors the equation for a commutator winding circuit is of the form

$$E_r = V - I_a R_a \text{ (V)}$$

Substituting for E_r in (7.9)

$$n = \frac{c}{2pZ} \times \frac{V + I_a R_a}{\Phi} = k \frac{V + I_a R_a}{\Phi} \tag{7.30}$$

Inspection of (7.30) shows that only two quantities can be varied in order to adjust the speed, namely:

- (a) the flux per pole Φ which is produced by the current I_f flowing through a concentrated coil winding,
- and (b) the supply voltage V, or more correctly the voltage which is applied across the commutator winding.

The methods of speed control are based on altering either the first quantity or the second or both simultaneously.

(a) Considering first the variation of the field current through the shunt connected winding, the equation (7.30) becomes:

$$n = k_1 \times \frac{1}{I_f} \tag{7.31}$$

where the symbol k_1 is a new constant. Since the supply voltage V is kept constant, $I_a R_a$ is small by comparison with V and can be neglected, and Φ the flux, may be assumed proportional to I_f.

Thus the speed of rotation is seen to be inversely proportional to the field current, which may be adjusted by means of the regulator R as shown Fig. 7.25.

The characteristic for separately and shunt excited motors shown in Fig. 7.18(a) can therefore be 'moved' up or down within the limits of the ohmic value of the rheostat R.

Figure 7.26 shows a family of curves of speed n plotted against torque. T_M for different settings of the regulator.

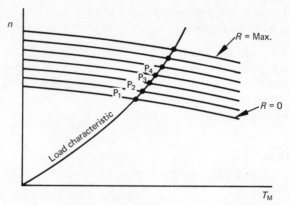

Fig. 7.26   Speed/torque characteristics for a shunt excited motor at different settings of R

It must be noted that the speed is lowest when the regulator R is set to zero, making the field current and therefore flux per pole a maximum. The upper limit for the speed is set by the maximum value of resistance of the regulator in series with the field, which makes I_f and consequently the flux a minimum.

The values of speed and torque which the motor imparts to the mechanical load can be found by plotting the Speed/torque curve of the load on the same graph with the motor's characteristics. Where the load characteristic intersects the motor curves the torques are exactly balanced and the machine works stably. The characteristic in Fig. 7.26 is that of the mechanical load, such as air fan and P_1, P_2 etc. are the operating points for different settings of speed controlling rheostat.

In case of the series excited motor, the armature current, determined by the mechanical load, would normally flow through the series field. A portion of that current however, can be diverted away from the series field coil by placing a variable resistor in parallel with it as shown in Fig. 7.27(a).

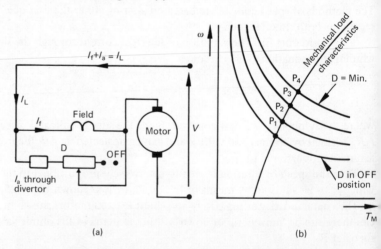

(a)   (b)

Fig. 7.27   (a) Speed control of series excited motor (b) Speed/torque characteristics for series excited motor at different settings of D

Hence the resistor D, called a divertor, reduces the current flowing through the field winding, which in turn reduces the magnetic flux and raises the speed.

Points P_1, P_2 etc. are the intersections of the d.c. series motor load characteristics and those of the mechanical load. They give the values of T_M and n at which the motor will drive a given load. Alternative method of controlling the speed of a series motor consists of splitting the series field coil into two halves, which may be connected in series or in parallel. This method does not give smooth speed control and only two specific characteristics from those shown in Fig. 7.27(b) can be chosen by a designer. The speed of compound motors is usually varied by adjustment of the shunt field current only, the series field current remaining unaltered by omitting the divertor from the circuit.

(b) Considering now the second available alternative that of varying applied voltage V to the armature winding the equation (7.30) becomes

$$n = k_2 V \tag{7.32}$$

where k_2 is yet another constant. The flux Φ is now kept constant and the $I_a R_a$ can be neglected as before.

Equation (7.32) indicates that the speed is directly proportional to V and, therefore, if means are provided for varying it smoothly from zero to any desired value, speed control over very wide limits can be obtained. The speeds include values which cannot be attained by the use of shunt field rheostat alone, because the resistance of the shunt field circuit cannot be reduced below the value of the winding itself.

The simplest method, but the most wasteful of electrical energy, for varying voltage across the commutator winding, is shown in Fig. 7.28(a).

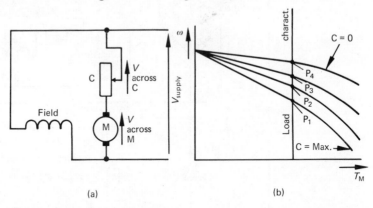

(a)   (b)

Fig. 7.28   D.C. motor with controller

The controller C is placed in series with the armature of the machine, much the same as a starter described earlier, except that it must be built to carry continuously the full armature current. Depending on its setting, greater or smaller p.d. is dropped across it and therefore the motor p.d. is reduced by the required amount from that of the supply voltage V. The speed is reduced accordingly. It is also clear from Fig. 7.28(b) that this method of speed control adversely affects the characteristics. The machine no longer operate as an almost constant speed motor.

However, at the present time, there are many efficient methods available for providing a variable d.c. voltage supply. These are shown diagrammatically in Fig. 7.29.

The box M represents any of the schemes which may be employed. In the Ward-Leonard system for instance, use is made of a motor-d.c. generator set. The motor may be a.c. or d.c. depending on the available supply.

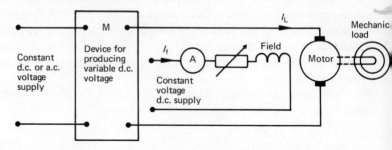

Fig. 7.29 General diagram of arrangement for controlling the speed of d.c. motors by adjustment of armature voltage

Alternatively, various forms of grid controlled mercury arc rectifiers, or more recently, thyristors and triacs, are utilized. As an example, one such scheme which utilises triacs is shown in Fig. 7.30.

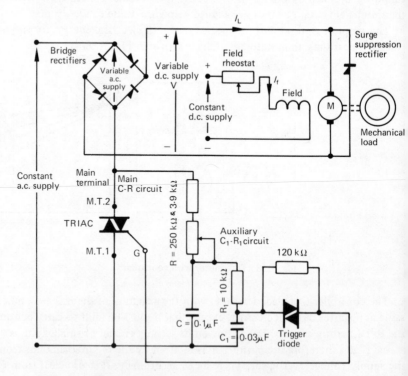

Fig. 7.30 Speed control of a small d.c. motor using triacs. (*Texas Instruments Ltd.*)

The triac is a semiconductor device which operates as a static switch. When a signal is applied to its gate G, the triac behaves as a switch the contacts of which are closed. However, as soon as an alternating current flowing through it, in either direction, falls to zero, the triac 'automatically' opens the circuit and has to be closed again by the impulse to its gate.

The gate signal is generated by a C–R network and applied to G via an auxiliary C_1–R_1 network and a trigger diode. By adjusting the variable resistor R the point during *each half cycle* at which the triac closes the circuit is controlled. Hence the average value of the a.c. voltage applied to the bridge rectifiers can be adjusted from zero to its maximum. The control of d.c. voltage V across the armature of the motor is thus achieved without appreciable loss of energy, and more importantly, without the 'droop' in its load characteristic.

Detailed descriptions however of this and other schemes lie outside the scope of this book.

QUESTION 11. The speed/torque characteristics for two different types of d.c. motors are given in the table below:

Torque (Nm)	20	40	100	200	400	600	700
Speed (rev/min)							
for shunt motor	1300	1295	1280	1250	1180	1080	1000
for series motor	1800	1400	1000	720	500	410	400

The torque of the mechanical load, which these two motors are to drive varies between the limits of 50 Nm and 350 Nm.

Find the limits between which the speed would vary if:

      (i) a shunt excited machine is used,

and   (ii) a series excited machine is used.

Calculate also the variation in power supplied by each motor at the lowest and highest limits.

ANSWER. The curves are plotted as shown in Fig. 7.31.

From the graph:
When the load torque is 50 Nm

          the speed of shunt motor = 1295 rev/min.
          the speed of series motor = 1300 rev/min.

When the load torque is 350 Nm

          the speed of shunt motor = 1195 rev/min.
          the speed of series motor = 530 rev/min.

Therefore speed change for shunt motor = 1295 − 1195 = 100 rev/min.
and   for series motor = 1300 − 530 = 770 rev/min.

Shunt Motor

$$\text{Maximum power supplied} = T\omega = \frac{350 \times 1195 \times 2\pi}{60} = 43\ 900 \text{ W}$$

$$\text{Minimum power supplied} = T\omega = \frac{50 \times 1295 \times 2\pi}{60} = 6800 \text{ W}$$

$$\text{Difference} = 37\ 100 \text{ W}$$

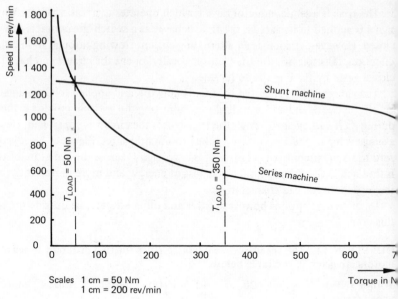

Fig. 7.31

Series Motor

$$\text{Maximum power supplied} = \frac{350 \times 530 \times 2\pi}{60} = 19\ 450\ \text{W}$$

$$\text{Minimum power supplied} = \frac{50 \times 1300 \times 2\pi}{60} = 6820\ \text{W}$$

$$\text{Difference} = 12\ 630\ \text{W}$$

Comparison of the figures shows that a shunt field motor is capable of delivering much greater power to a larger load because of its ability to maintain almost constant speed. The series excited motor on the other hand, will only provide higher torque at a much lower speed and reduced power. The above example clearly illustrates the difference in characteristic operation of the two machines.

7.20   Present day use of d.c. machines

GENERATORS

The use of d.c. machines as primary generators of electrical energy is very limited since the bulk of energy is generated in the form of alternating current. In general, therefore, d.c. generators are confined to supplying the excitation of medium range a.c. generators or to converting a.c. to d.c. for industrial applications.

Although in industry itself direct current is widely used for electrolytic processes, for variable speed motor drives and to a small extent in welding processes, nevertheless the trend is away from d.c. generators and towards static rectifiers.

Even in the steel industry, semiconductor rectifiers are slowly replacing d.c. generators as sources of supply for mill applications. The quality of direct current, that is freedom from ripple, is better when obtained from a d.c. machine. But the d.c. generator with its rotating commutator requires expert maintenance and

operation and regular replacement of worn parts. These factors are principally
responsible for its declining use.

MOTORS

Direct-current motors still lead the field in such application as variable-speed devices
and where severe torque variations occur. In the future the variable-frequency a.c.
drives may replace the d.c. drives for some applications, but for many applications,
d.c. motors will remain most suitable.

As stated above, d.c. is no longer available from the national network, hence the
most common practice today is to have an individual semiconductor rectifier for each
d.c. motor, or for a group of motors. Alternatively a.c. motor-d.c. generator sets may
still be used, but for the most part these have been replaced by the rectifiers.

The rectifier and the controls of a d.c. motor are often designed as a packaged
drive and fitted in the same enclosure.

The applications of the three types of d.c. motors are summarised below:

(a) Series excited motors are used where high starting torque is required and speed
    can vary: traction, cranes, car crashers.
(b) Shunt excited motors are used where constant speed is required
    and starting conditions are not severe: fans, blowers, centrifugal pumps,
    conveyors, wood and metal working machines, lifts.
(c) Compound excited motors are used where high starting torque and fairly
    constant speed is required: plunger pumps, presses, shears, bending rolls,
    geared lifts, conveyors and hoists.

SUMMARY

1. A d.c. machine is defined as having a concentrated coil winding on the stator and a
commutator on the rotor both carrying direct currents.

2. Power conversion equation of the machine's air-gap:

$$E_r I_a = T\omega. \tag{7.1}$$

where E_r = average e.m.f. of rotation induced in the commutator winding
$\quad\quad I_a$ = total armature current
$\quad\quad T$ = average torque at which power conversion takes place
and $\quad \omega$ = constant speed of rotation of the machine.

3. The 'armature reaction' is the name given to the effect of the current in the
commutator winding on the magnetic flux produced by the concentrated coil winding.

4. Commutating poles are placed in-between the main poles in order to neutralise
the effects of armature reaction in brush region and eliminate sparking at brushes.
Ampere-turns per pole on each compole to achieve this are equal to

$$\frac{1}{2} \cdot \frac{Z}{2p} \cdot \frac{I_a}{c} \text{ (AT)} \tag{7.2}$$

5. E.m.f. of rotation $= \dfrac{pZ}{\pi c} \times \Phi\omega = 2p\Phi\dfrac{Z}{c}n \text{ (V)}$ $\tag{7.5/6}$

where Z = Total number of armature conductors

                               p = No. of pairs of poles

                               c = 2 for wave winding

                                  = $2p$ for lap winding

                               Φ = flux per pole in Wb.

                               ω = speed of rotation in rad/s. and n = speed in rev/s.

6.
$$\text{Torque} = \frac{pZ}{\pi c} \times \Phi \times I_a \text{ (Nm)} \tag{7.7}$$

where I_a = Armature current

7.
$$\text{Speed of rotation } n = \frac{c}{2pZ} \times \frac{E_r}{\Phi} \text{ (rev/s.)} \tag{7.9}$$

8. Two winding d.c. machines may be interconnected to give:
   (i)   separately excited machine
   (ii) shunt excited machine
   (iii) series excited machine.

Three winding d.c. machines may be interconnected to give:

       (i)   commulative compound connections either short shunt or long shunt,
   or (ii) differential compound connections either short shunt or long shunt.

9. For separate and shunt connections
$$V_f = I_f R_f \tag{7.10}$$
$$V = E_r \pm I_a R_a \tag{7.15}$$

where I_a is +ve for the motor and −ve for the generator

and
$$VI_a = T\omega \pm I_a^2 R_a \tag{7.16}$$

where $I_a^2 R$ is +ve for the motor and −ve for the generator

For series connections
$$V = E_r \pm I(R_a + R_f) \tag{7.23}$$
and
$$VI = T\omega \pm I^2(R_a + R_f) \tag{7.24}$$

10. Power equations including rotational losses C are for the motor:
$$T\omega = E_r I_a = T_M \omega + C \tag{7.25}$$

and for the generator
$$T\omega = E_r I_a = T_P \omega - C \tag{7.26}$$

where T_M = net torque of the motor

                               T_P = prime mover torque

                               T = internal torque, in the d.c. machine at which electro-
                                      mechanical power conversion takes place.

11. Rotational Losses are due to
   (i)   friction in bearings
   (ii) windage
   (iii) core loss

12. Characteristics are graphs which relate: terminal voltage V to load current I_L when the machine works as a generator and torque T_M to the speed of rotation ω when the machine works as a motor.

13. Generators are controlled by:

(i) adjustment of driving speed of the mechanical prime-mover

(ii) adjustment of e.m.f. of rotation, which determines the output voltage V.

The second is done by altering the flux per pole Φ, produced by field current I_f, by means of a rheostat connected in series with the exciting winding.

14. The control of motors involves (i) starting from rest and (ii) the ability to alter the speed of rotation. Starting is accomplished with the help of the STARTER. Speed control is achieved either by varying the field current through exciting winding:

$$n = k_1 \frac{1}{I_f} \tag{7.31}$$

or by varying the applied voltage V to the commutator winding of the rotor:

$$n = k_2 V \tag{7.32}$$

EXERCISES

1. Define a d.c. machine in terms of windings and the currents carried by them. Explain simply why this definition is sufficient to identify the machine.

2. Using the equation

$$e = Blu \text{ (V)}$$

and

$$f = Bli \text{ (N)}$$

explain briefly the principle of operation of a d.c. machine working as (i) a generator; (ii) a motor.

3. Sketch the waveform of flux density B in the two pole d.c. machine's air-gap due to

(i) Field current flowing in the stator concentrated coil winding alone;

(ii) Armature current flowing in the rotor commutator winding alone;

(iii) Combination of (i) and (ii).

Using the above, explain what is meant by armature reaction, and what are its chief effects.

4. 'E.m.f. of rotation E_r occurs in a d.c. machine irrespective of whether it operates as a generator or a motor'.

Justify this statement and explain how E_r occurs under each mode of operation.

5. Describe the test to obtain the Open Circuit characteristic of a 240-V, 1500 rev/min. d.c. machine, whose field coil resistance is 120 Ω. The description should include:

(i) the equation of the e.m.f. of rotation in terms of number of poles, conductors per parallel path, flux per pole and speed of rotation.

(ii) a circuit diagram with the list of instruments, rheostats etc.

(iii) calculation of range for the field ammeter, the voltmeter and the field rheostat to vary the field current down to 0·1 A.

(iv) expected shape of the graph.

6. (a) State the number of parallel paths in the armature of a 4-pole d.c. generator when it is:

(i) lap-wound, and

(ii) wave-wound.

(b) Sketch the circuit diagrams of five different methods of field excitation for d.c. generators, and name the methods.

(c) An 8-pole generator has a flux per pole of 0·051 Wb. The lap wound armature has 1088 conductors and rotates at 400 rev/min. Calculate the e.m.f. generated. (Derive formula used.)

Answer: 370 V.

7. An eight-pole lap connected commutator winding has 880 conductors, and carries a current of 6 A per parallel path. When the winding is placed inside the stator, which produces a useful flux per pole of 0·06 Wb, calculate the torque which the rotor will experience in (i) Nm, (ii) lbf ft.

Answer: 807 Nm,   595 lbf ft.

8.   (i)   What is the function of the compoles in d.c. machines?
    (ii) Why are they used for d.c. machines operating as motors as well as generators?
    (iii) What type of winding is placed on a compole, and how is it connected to other windings of the machine?
    (iv) What current is used to supply the commutating winding?
    (v)   Calculate the ampere-turns and the number of turns necessary to counter-balance the effect of armature for the machine in Question 7.

Answer: (v) 330,   7 approx.

9. Describe how to set the brushes of a 2-pole d.c. motor on the geometric neutral axis. Explain why the method gives the required result.

What other adjustments should be made to the brushgear in order to give good commutation?   (C. & G.L.I. ET.PtII 1969)

10. The following particulars refer to a d.c. machine connected to 400-V d.c. busbars:

Number of poles	6
Flux per pole	30 mWb
Number of armature conductors connected in simplex lap	864
Armature resistance	0·2 Ω
Shunt field resistance	200 Ω.

  (i) When the machine is operated as a shunt excited motor taking a current of 50 A from the bus-bars, calculate the speed and torque it imparts to the mechanical load.
  (ii) What would be the torque and speed of the prime-mover if the machine is driven as a d.c. generator and delivers a current of 50 A to the d.c. bus-bars? Comment on your results. Neglect Windage, Friction and Core Losses.

Answer: (i) 905 rev/min, 197 Nm;   (ii) 951 rev/min, 215 Nm.

11. Draw the equivalent circuits indicating direction of currents, e.m.f.'s and potential differences for a shunt wound d.c. machine operating as:

  (i) a generator,
  (ii) a motor.

Explain what each element shown represents in the actual machine and deduce equations relating currents and e.m.f.s for each case.

Convert these expressions into power equations and use them to explain how such a machine converts electrical energy to mechanical energy and vice versa.

12. A compound wound d.c. motor has four pairs of terminals connected to the brushes, the commutating poles, the shunt winding and the series winding respectively: the terminals are unmarked.

Describe how you would identify the terminals, and find the correct connections for the machine to run in a given direction of rotation as a cumulative compound motor.   (C. & G.L.I., ET Pt.II, 1968)

13. A 'short-shunt' compound wound d.c. machine is connected to a 440-V d.c. supply and run as a motor taking a current of 40 A. The machine's particulars are:

Armature resistance	$0.2 \ \Omega$
Series field resistance	$0.3 \ \Omega$
Shunt field resistance	$200 \ \Omega$

(a) If the speed of rotation is 1500 rev/min, calculate the rotational e.m.f. generated and gross mechanical torque.

(b) If the rotational losses amount to 500 W, calculate all the resistive losses in the machine and find its efficiency.

Answer: (a) 420·4 V, 322 Nm.   (b) Total resistive losses 1682 W
Efficiency 0·876.

14. A 'long-shunt' compound d.c. machine is connected to 400-V d.c. busbars. It is driven as a generator by a mechanical prime-mover and feeds 40 kW into the d.c. network. The machine's particulars are:

Armature resistance	$0.1 \ \Omega$
Series field resistance	$0.2 \ \Omega$
Shunt field resistance	$100 \ \Omega$

If the speed of rotation is 1000 rev/min. calculate the rotational e.m.f. generated and the internal torque.

Assuming the rotational losses in the machine amount to 1400 W, what is the driving torque of the prime-mover and the efficiency of the generator?

Draw a complete circuit diagram marking all the currents and voltages on it.

Answer: 431·2 V,   430 Nm,   442 Nm,   0·866.

15. A 22 400-W shunt excited motor is run unloaded from a 240 V d.c. supply and takes 5 A. $R_a = 0.08 \ \Omega$ and $R_f = 120 \ \Omega$.

Assuming that the rotational loss is the same on no load as at full load, calculate its value and also the efficiency of the motor when fully loaded.

What is the nett Torque T_M produced by the motor on full load if the speed is 900 rev/min?

Answer: 719·3 W,   0·916,   237 Nm.

16. Sketch the shape of a load characteristic (terminal voltage plotted against load current) for

    (i) Separately excited d.c. generator
    (ii) Shunt excited d.c. generator
    (iii) Series excited d.c. generator
    (iv) Compound excited d.c. generator.

Justify each shape by using appropriate equations.

17. Sketch the shape of a load characteristic (speed torque) for

    (i) Separately excited d.c. motor
    (ii) Shunt excited d.c. motor,
    (iii) Series excited d.c. motor,
    (iv) Compound excited d.c. motor.

Justify each shape by reference to appropriate equation.

18. Sketch the speed-torque and current-torque characteristic curves of a d.c. shunt-connected motor, and explain in detail the reasons for their shape.

A d.c. series motor takes 30 A from a constant voltage source when developing full load torque at 1500 rev/min. Neglecting saturation and losses, estimate the current the motor will take, and the speed at which it will run, when developing $\frac{1}{4}$ full load torque.

Answer: 15 A,   3000 rev/min.   (C. & G.L.I., ET Pt. II, 1967)

19. The load characteristic for a shunt excited d.c. generator was obtained experimentally and is given below:

Terminal Voltage (V)	300	290	285	270	250	230	190	150	110	60	40	25	0
Load current I_L (A)	0	20	30	40	60	70	80	82	80	70	60	50	30

Plot the graph of V against I_L and determine the value of voltage, current and the power the generator will supply to the load of

    (i) 4 Ω resistance
    (ii) 15 Ω resistance
    (iii) 0 Ω resistance (i.e. short circuit)

Answer: (i) 15 A, 246 V;   (ii) 19·5 A, 292·5 V;   (iii) 30 A, 0 V.

20. The open-circuit characteristic of a d.c. shunt generator driven at 400 rev/min is given by the following data:

Field current (A)	1·0	2·0	3·0	4·0	5·0	6·0	7·0	8·0	9·0
E.m.f. (V)	60	110	155	186	212	230	246	260	271

Find by construction:

    (a) the e.m.f. to which the machine will excite when the field circuit resistance is 34 Ω.
    (b) the critical value of the shunt field circuit resistance at 400 rev/min
    (c) the additional resistance required in the field circuit to reduce the e.m.f. to 220 V at 400 rev/min
    (d) the critical speed when the field circuit resistance is 34 Ω.
    (e) the lowest speed at which an e.m.f. of 225 V can be obtained with a field circuit resistance of 34 Ω.
    Neglect the volt-drop in the armature due to the shunt field current.

Answer: (a) 248 V,   (b) 60 Ω,   (c) 3·2 Ω,   (d) 227 rev/min., (e) 350 rev/min.

21. (a) Explain why it is usually necessary to use a starter for a d.c. motor.

(b) Draw a circuit diagram showing a shunt-connected motor fitted with the starter complete with no-volt and over-load releases. Explain fully the operation of the starter.

22. The speed/torque characteristics of a shunt excited d.c. motor and the mechanical load which it is to drive are given below:

Torque (Nm)	20	40	100	200	400	600	700
Speed of shunt motor (rev/min)	1300	1295	1280	1250	1180	1080	1000
Speed of load (rev/min)	20		100	240	740	1800	

Plot the graphs and find the speed and torque at which the motor drives the load.

If the speed of the load rises by 60 rev/min find the maximum retarding torque and power which will return the system to its stable operating point.

Answers: 530 Nm,   1120 rev/min,   10 Nm,   62·8 Wp.

23. (a) Explain how the speed of a d.c. shunt motor may be controlled by variation of the shunt field circuit resistance.

(b) A 246-V d.c. shunt motor runs at a speed of 800 rev/min with an armature current of 18 A. If the armature circuit resistance is 0·1 Ω calculate the value of:

    (i) the new armature current,

    (ii) the new speed,

        if the torque is doubled and the field weakened by 10%.

Answers: 40 A,   880 rev/min.

24. A 240-V shunt motor has an armature resistance of 0·5 Ω and a field resistance of 240 Ω. The armature current is 20 A at 600 rev/min when driving a load whose torque is constant. Calculate:

    (a) the value of resistor to be placed in series with the armature to lower the speed from 600 rev/min to 400 rev/min.

    (b) the value of the series resistor to be inserted in the field to raise the speed from 600 rev/min to 800 rev/min.

        Assume field flux proportional to field current.

Answers: 4·33 Ω, 80 Ω.

$$-0.5$$
$$= 3.83\,\Omega$$

Assuming E_b constant
otherwise 67·6 Ω

Three-phase induction machines

This chapter describes the theory of a three-phase induction machine which may be operated as (i) a motor, (ii) a generator, and (iii) an induction regulator.

The explanation of its action as (i) and (ii) is based on an electromechanical power conversion relationship $T\omega = e_r i$. Although the emphasis placed on reversibility of the machine indicates equal importance of the motor and generator action, nevertheless in industrial application an induction machine is almost universally employed as a motor. Its use as a generator or an induction regulator is minimal by comparison.

Single phase induction motors are described in Chapter 11.

8.1 Definition of induction machines

An induction machine may be defined as having a phase type winding on the STATOR and a phase or cage type winding on the ROTOR.

Both windings carry alternating currents, either three-phase or single phase depending on the supply for which they are constructed. It must be stressed that the windings are different in each case and the machine made for one supply cannot be used on the other.

In this chapter only the three-phase type is described being the most commonly used in industrial applications.

Single phase machines, which are usually very small, (generally less than 1 kW) are dealt with in Chapter 11 'Small Electric Motors'.

8.2 Construction of typical three-phase induction machines

A large induction machine is illustrated in Chapter 4. A typical small rotating machine. is shown here in Fig. 8.1.

The illustration shows a common stator with two rotors. The first marked 1 is a cage type whilst the second marked 2 is a wound type carrying a phase winding brought out to three slip rings.

Figure 8.2 shows an induction machine which is used as a static transformer. Its construction is the same as that of the machine in Fig. 8.1 where the rotor is marked 2, but the rotor slip rings are omitted. This machine is also termed an induction regulator.

Fig. 8.1   Stator and two rotors of a small 3-ph induction motor.
(*N.E.C.O. Ltd.*)

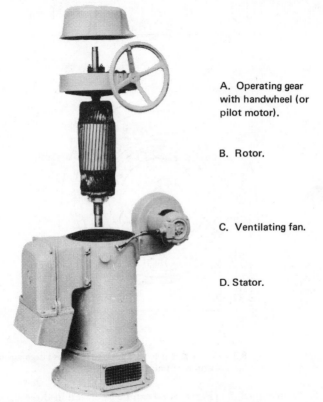

A. Operating gear
with handwheel (or
pilot motor).

B. Rotor.

C. Ventilating fan.

D. Stator.

Fig. 8.2   Induction regulator, hand operated and with independent cooling
fan. (*Laurence, Scott & Electromotors Ltd.*)

8.3   Three methods of operating an induction machine

The two basic types of three-phase induction machine defined in paragraph 8.1 are
shown in cross-section in Figs. 8.3(a) and (b).

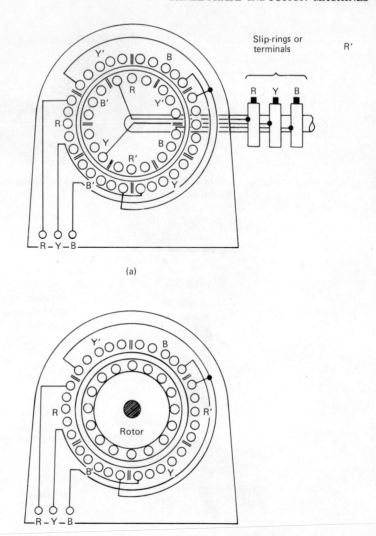

(a)

(b)

Fig. 8.3   (a) Wound rotor machine (b) Squirrel cage machine. Cross-sections of induction machines

The wound rotor machine (Fig. 8.3(a)) can be operated in three different ways as follows:

(i)   as an induction motor to drive a mechanical load; when the stator winding is fed from a three-phase supply and the rotor winding is short-circuited,

(ii) as an induction generator to supply power to an a.c. network; when the short-circuited rotor is driven by the prime-mover and the stator winding is connected to the network, and

(iii) as an induction transformer (regulator) to supply an electrical load at required potential and phase shift; when one winding is connected to the supply and the other to the load. In this case the rotor remains stationary and no electro-mechanical power conversion takes place. Furthermore, the sliprings are replaced by fixed terminals and flexible connections.

The squirrel cage machine illustrated in Fig. 8.3(b) can only be used as a motor or a generator, but not as an induction transformer. This is obvious since the cage winding has no external connections.

8.4   Simple explanation of operation

THE INDUCTION MOTOR

In order to understand the operation of the motor, consider first the arrangement shown in Fig. 8.4, where a two-pole stator *and* a rotor carrying a single short-circuited coil can both rotate independently.

If the stator is driven by a mechanical torque, T_S, in an anticlockwise direction with the speed ω_S, then the magnetic flux Φ produced by the two salient poles 'cuts' both sides of the coil, inducing in it an instantaneous e.m.f. e_R according to the relationship:

$$e_R = Bl\omega_S r \text{ (V)}$$

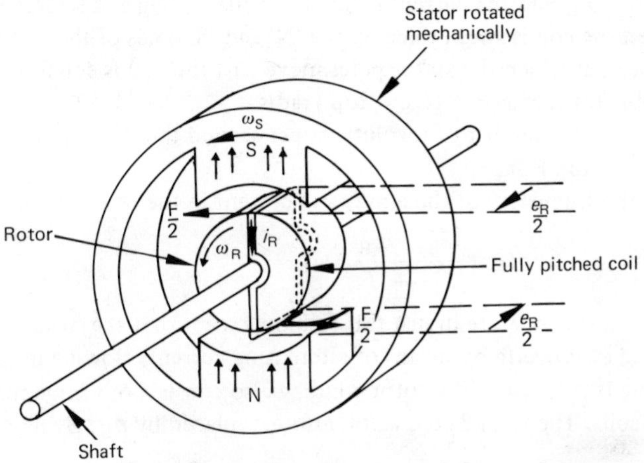

Fig. 8.4   Shaft with free moving stator and rotor

The direction of e_R shown in Fig. 8.4 is obtained by Fleming's R.H. Rule. It must be remembered, however, that in this case the conductor motion is clockwise with respect to the magnetic flux, if the latter is thought of as fixed.

The e.m.f. e_R produces the current i_R around the coil in the direction shown by the arrow, and the mechanical power supplied to the stator is converted to electrical power in the initially stationary rotor coil, i.e.,

$$T_S\omega_S \text{ (Nm/s)} = e_R i_R \text{ (W)}$$

The sides of the coil are subjected to magnetically generated forces, according to the equation

$$F = Bli \text{ (N)}$$

The direction of each force is found by applying Fleming's L.H. Rule. It is clear that these forces tend to turn the coil in the same direction as the motion of the stator, that is, anticlockwise. Since the rotor is not fixed, it follows the stator and revolves anticlockwise with the speed ω_R. The rotor in turn imparts to the mechanical load a torque T_R. The electrical power now generated in the coil is reduced by the amount $T_R \omega_R$ because e_R is small due to the reduction of the flux cutting speed from ω_S to $(\omega_S - \omega_R)$ i.e.,

$$e_R = Bl(\omega_S - \omega_R)r \text{ (V)}.$$

The new power equation is:

$$T_S \omega_S = e_R i_R + T_R \omega_R$$

Under these conditions most of the mechanical power intake $T_S \omega_S$ is transferred directly to the mechanical load via the magnetic field. The balance given by $e_R i_R$ is the electrical power lost in the resistance of the rotating coil. However without the presence of e_R and i_R, there would be no torque on the rotor coil and the system could not operate. Furthermore, the speed of the rotor ω_R must always be smaller than ω_S, because the generation of e_R depends on the cutting of magnetic flux and this necessitates the difference in speed between the coil and the stator.

The e.m.f. e_R and the current i_R are both alternating and undergo one complete cycle when the coil is swept once by the 'N' and 'S' poles of the stator.

The speed at which the stator poles move past the coil is equal to the difference of stator and rotor speeds, i.e. $(\omega_S - \omega_R)$ rad/s.

Therefore, the number of revolutions per second the stator poles move past the coil is: $(\omega_S - \omega_R)/2\pi$.

Hence the frequency of the e.m.f. and current in the rotor coil is:

$$f_R = \frac{\omega_S - \omega_R}{2\pi} \text{ (Hz)} \tag{8.1}$$

Induction motors operate in just this fashion except that the 'rotating stator' effect is produced *electrically* by means of alternating currents flowing in phase windings placed on a fixed stator. The rotor winding also consists of a large number of short-circuited coils. The term $T_S \omega_S$ is, therefore, replaced by p_E, giving equation:

$$p_E \text{ (W)} = e_R i_R \text{ (W)} + T_R \omega_R \text{ (Nm/s)} \tag{8.2}$$

Where p_E is the instantaneous electrical power input to the stator. All losses, except resistive loss in the rotor coil, are neglected in the above.

In the two induction machines shown in Figs. 8.3(a) and (b), the stators with three-phase windings are shown in place of the mechanically rotating salient poles illustrated in Fig. 8.4. The stators produce the same effect as the two poles, i.e. they generate a rotating magnetic field pattern which is constant in magnitude and has a uniform sinusoidal distribution of flux density in a uniform air-gap. (See Chapter 6 paragraph 6.13.)

THE INDUCTION GENERATOR

Consider either of the machines, shown in Fig. 8.3(a) or Fig. 8.3(b). The operation of an induction machine as a generator requires that its stator winding is connected to the three-phase supply in order to establish a rotating magnetic field, and that its short-circuited rotor is driven at a speed higher than that at which the field revolves.

The speed of the magnetic field is determined by the frequency of the supply and is deduced in the next paragraph.

When the rotor revolves faster than the magnetic field, then its conductors 'cut' the flux lines in *opposite sense* to that in the motor case.

The e.m.f.s and the currents in the rotor winding are thus 180° out of phase with the e.m.f.s induced in it during motor operation. By transformer action *the current* in the stator winding must, therefore, also be 180° out of phase with the motor case.

The stator e.m.f.s and currents have now the same sense and the power flow through the machine is reversed.

The mechanical power supplied by the prime-mover is converted to electrical power through the magnetic link and fed back into an a.c. network.

It must be stressed that an induction machine does not generate power if it is operated in isolation, i.e., not connected to the supply. This is because neither the stator not the rotor winding is excited electrically and, therefore, no magnetic field exists in the air-gap.

It might be added that under special conditions when capacitors are connected across the stator terminals, the machine may be made to operate as a generator. Such a case, however, is not considered here.

INDUCTION TRANSFORMER (REGULATOR)

Consider now the machine shown in Fig. 8.3(a) but with its slippings replaced by flexible connections.

Such a machine is sometimes used as a three-phase transformer usually termed an induction regulator. Its purpose is to transform the p.d. and to enable the phase angle between the supply and the p.d. of the load to be varied.

When the stator winding is connected to the supply a rotating magnetic field is established in the air-gap. This field, by virtue of its rotation, cuts the rotor windings and induces three-phase e.m.f.s.

If the rotor is now connected to a load, electrical power is delivered. The rotor is not allowed to revolve but its position relative to the stator can be adjusted manually. The phase angle between the input and the output p.d.s can thus be controlled.

The value of the p.d.s depends on the turns ratio between the stator and the rotor windings and once the machine is built this cannot be altered. However by interconnecting the windings it is possible to make the machine into a variable voltage device as shown in paragraph 8.10.

The machine operates equally well if the supply and the load connections are interchanged between the stator and the rotor windings.

8.5   Speed of magnetic field

A stator with three-phase distributed windings, shown in Fig. 6.16 in Chapter 6 and supplied with alternating currents, is *equivalent* to *one pair of poles*. The poles revolve through 2π radians for one complete current cycle. Therefore, if the three-phase currents undergo f cycles per second, that is if their frequency is f hertz, then the speed of the magnetic field is f revolutions per second or $2\pi f$ radians per second. Hence the speed of a magnetic field ω_S is given by

$$\omega_S = 2\pi f \text{ (rad/s.)}$$

Consider now the stator shown in Fig. 8.5 where the number of conductor 'bands' is doubled and, therefore, each 'band' occupies 30° instead of 60°.

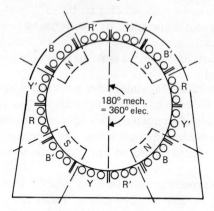

Fig. 8.5   Stator with six coils equivalent to two pairs of salient poles

Since the six 'bands' produce the effect equal to one pair of poles, twelve 'bands' produce the effect equal to two pairs of poles, as shown by the dotted lines in Fig. 8.5.

Furthermore, one input current cycle causes each pair of poles to travel through the portion of the stator's circumference occupied by the six 'bands' producing them. In this case, therefore, one cycle causes half a revolution, as the six 'bands' are accommodated in one half of the stator's circumference.

Therefore, if f is the frequency of the currents, then the speed of the two pairs of poles,

$$\omega_S = \frac{2\pi f}{2} \text{ (rad/s.)}$$

Trebling the number of conductor bands to the total of 18 produces three equivalent pairs of poles and the speed:

$$\omega_S = \frac{2\pi f}{3} \text{ (rad/s.)}$$

Hence if p is the number of equivalent pairs of poles, then the speed of the magnetic field is given by:

$$\omega_S = \frac{2\pi f}{p} \text{ (rad/s.)} \tag{8.3}$$

or

$$n_S = \frac{2\pi f}{p} \times \frac{1}{2\pi} = \frac{f}{p} \text{ (rev/s.)} \tag{8.4}$$

where n_S = speed of magnetic field in revolutions per second.

8.6   Slip speed and slip

If the rotor carrying short-circuited coils is placed inside the stator, which produces a rotating magnetic field at $\omega_S = 2\pi f/p$ (rad/s), then the rotor will follow the field at a reduced speed ω_R (rad/s), as described in paragraph 8.4.

The difference between the two speeds is termed the 'slip speed', since the rotor may be thought of as slipping backwards through the magnetic field. Thus

$$\text{slip speed} = (\omega_S - \omega_R) \text{ rad/s} = (n_S - n_R) \text{ rev/s}.$$

An additional factor termed 'SLIP' is used in the theory of induction machines and is defined as the ratio of slip speed to the speed of the magnetic field. The ratio is denoted as s, i.e.,

$$s = \frac{\omega_S - \omega_R}{\omega_S} = \frac{n_S - n_R}{n_S} \text{ p.u.} \qquad (8.5)$$

Slip has no dimensions and gives the 'per unit' difference between the slip speed and the speed of the magnetic field in the air-gap.

When the machine operates as a motor, ω_S is greater than ω_R and the slip 's' is positive. When the machine operates as a generator, then its rotor must be driven at a speed greater than ω_S and the slip is negative.

Typical values of slip for industrial machines are as follows:

Motor: free running at no load   about + 0·005 p.u.
                   at full load   + 0·03 to + 0·07 p.u.
Generator: at full load   − 0·03 to − 0·05 p.u.

QUESTION 1. Find the speed of rotation in rev/s and rev/min. for a three-phase, 4-pole induction machine operating from 50 Hz supply, at:

(i)   no load slip of + 0·005 p.u.
(ii) full load slip of + 0·06 p.u. (motor)
(iii) full load slip of − 0·04 p.u. (generator)

ANSWER. The speed of magnetic field from equation (8.4).

$$n_S = \frac{f}{p} = \frac{50}{2} = 25 \text{ rev/s or } 1500 \text{ rev/min.}$$

transforming equation (8.5)

$$n_R = n_S - sn_S = 25 - 25s$$

(i) $n_R = 25 - 25 \times 0·005 = 24·875$ rev/s. or 1492·5 rev/min.
(ii) $n_R = 25 - 25 \times 0·06 = 23·5$ rev/s or 1410 rev/min.
(iii) $n_R = 25 - 25 \times (−0·04) = 26$ rev/s or 1560 rev/min.

8.7   Frequencies of currents and e.m.f.s in an induction machine

The frequency of currents and e.m.f.s in the stator winding must be the same as the frequency of the supply. It is related to the speed of the magnetic field by an equation:

$$\omega_S = \frac{2\pi f}{p}$$

therefore stator frequency

$$f = \frac{p\omega_S}{2\pi} \text{ (Hz)} \tag{8.6}$$

The frequency in the rotor windings is, however, variable and was shown in paragraph 8.4 to depend on the difference between the rotor speed and the speed of the magnetic field.

For a two pole arrangement it is given by expression (8.1) as:

$$f_R = \frac{1}{2\pi}(\omega_S - \omega_R)$$

If the machine has 'p' pairs of poles, then the equation must be multiplied by 'p', i.e.,

$$f_R = \frac{p}{2\pi}(\omega_S - \omega_R) = \frac{p}{2\pi} s\omega_S = \frac{p}{2\pi} s \frac{2\pi f}{p} = sf$$

Since from equation (8.5)

$$s\omega_S = \omega_S - \omega_R$$

and

$$\omega_S = \frac{2\pi f}{p} \text{ by equation (8.3)}.$$

Thus

$$f_R = sf \text{ (Hz)} \tag{8.7}$$

It is clear from the above that f_R varies from a value equal to that of the supply frequency at standstill ($s = 1$) to zero ($s = 0$), when the rotor is driven by a mechanical prime-mover at synchronous speed ω_S. When the rotor speed is raised above ω_S the frequency again begins to increase above zero. Its value will be negative, since s is negative. The minus sign, however, should be interpreted merely as an indication of rotor conductors moving faster than the rotating magnetic field.

8.8   Circuit representation of an induction machine

The induction machine may be drawn in a simplified form as shown in Fig. 8.6.

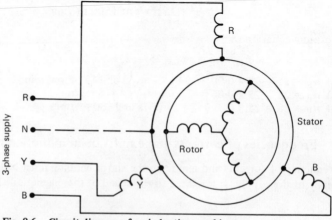

Fig. 8.6   Circuit diagram of an induction machine

The stator winding is represented by three coils one for each phase. The coils are shown to be connected in STAR although in practice they are often in DELTA configuration. Similarly the rotor winding is represented by three coils joined in STAR and drawn 90° with respect to those on the stator. Again the coils may be either in star or in delta connection.

For purposes of analysis it is helpful to represent each winding by three elements connected in series, each denoting one quantity as follows:

   (i) an e.m.f. of rotation induced in the winding,
   (ii) an ohmic resistance of the winding,

and   (iii) a self inductance of the winding due to that portion of a flux which links *only* the winding conductors themselves. (Fig. 8.7).

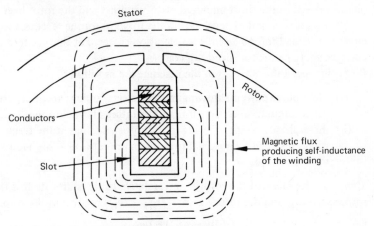

Fig. 8.7   **Self-inductance of the phase winding**

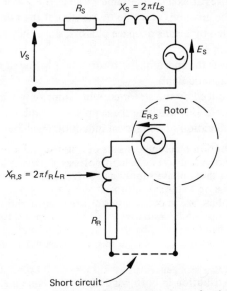

Fig. 8.8   **Circuit representing one phase of an induction machine**

As all three-phases of the machine are identical it is only necessary to consider one (Fig. 8.8).

In the diagram suffix S refers to the stator, whereas suffix R to the rotor quantities. Second suffix s indicates that the machine is working at slip s. For the machine at standstill o is used in place of s. Thus $E_{R,s}$ denotes the phase e.m.f. of the rotor winding when the machine operates at slip s.

8.9   Operation of a practical induction machine – general considerations

As soon as the stator windings are connected to a three-phase supply, a sinusoidal rotating magnetic field is established in the machine's air-gap. This field cuts the rotor conductors and induces three-phase e.m.f.s. These e.m.f.s in turn establish three-phase currents in the short-circuited rotor windings and the rotor begins to revolve.

After a short period of acceleration the forward torque reaches a value which is equal to the load torque and the acceleration ceases. Let this occur at a slip s corresponding to a rotor speed ω_R (rads/s.)

At this speed the currents in the machine are as follows:

(i) three-phase currents flowing in the stator windings at supply frequency f (Hz) which produce a sinusoidal m.m.f. rotating at synchronous speed ω_S

(ii) three-phase currents flowing in the rotor windings at the frequency $f_R = sf$ (Hz), which produce a sinusoidal rotating m.m.f. at $(\omega_S - \omega_R)$ rad/s relative to the rotor's surface.

Therefore the speed of the rotor m.m.f. relative to the stator itself is the sum of the rotor speed ω_R and the rotor's m.m.f. $(\omega_S - \omega_R)$ relative to the rotor surface.

i.e., $$\omega_R + (\omega_S - \omega_R) = \omega_S$$

Thus the rotor m.m.f. revolves at exactly the same speed as the stator m.m.f. The rotating magnetic flux in the air-gap is now due to the sum of the two m.m.f.s, which being sinusoidal can be added by means of space phasors, since they revolve in unison and always at synchronous speed.

The above is true whatever the speed of the rotor; either below or above that of the speed of the stator's magnetic field.

The rotating field is produced by the stator currents alone only when the rotor windings are on open circuit or on switching the supply to an induction machine. Question 2 illustrates the operation of a practical induction machine.

QUESTION 2. (i) An induction machine consists of a stator and a rotor both carrying three-phase windings equivalent to two salient poles as shown in Fig. 8.9. The stator winding is fed from a three-phase supply at 50 Hz whilst the rotor winding is short circuited. The machine operates as a motor at slip $s = 0.04$ p.u. At an instant when the currents in RR' phase are maximum in both the stator and the rotor windings, the rotor m.m.f. F_R is $5\pi/6$ radians (150°) behind the stator m.m.f. F_S.

If the maximum value of m.m.f.'s due to stator and rotor windings is 120 A and 110 A respectively, find the resultant air-gap m.m.f. F_A. Assume an anticlockwise rotation of the motor.

(ii) If the machine operates as a generator at slip $s = -0.04$ p.u., draw its cross-section showing current distribution in both windings at the same instant as for the motor case.

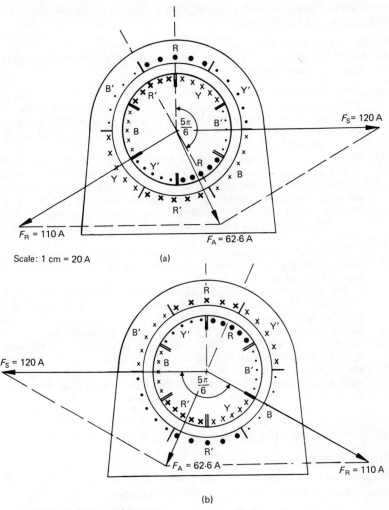

Scale: 1 cm = 20 A   (a)

Fig. 8.9   (a) Motor (b) Generator

ANSWER

(*i*) *Motor operation*

Stator m.m.f. F_S rotates at the same speed as the frequency of the supply, i.e., 50 rev/s.

Rotor revolves at $n_R = n_S - sn_S = 50 - 0.04 \times 50 = 48$ rev/s.

Frequency of currents in the rotor is $f_R = sf = 0.04 \times 50 = 2$ Hz. Hence the rotor m.m.f. F_R revolves at 2 rev/s with respect to the rotor itself.

Speed of F_R with respect to the stator = 48 + 2 = 50 rev/s, i.e. the same as for stator m.m.f.

Since the relative position of the rotor m.m.f. with respect to the stator m.m.f. is known and since it remains constant at a given load, therefore F_S and F_R can be added by phasor addition, i.e.,

$$F_A = F_S + F_R$$

From the phasor diagram in Fig. 8.9(a) $F_A = \underline{62 \cdot 6 \, A}$, and is responsible for producing the air-gap flux of the machine.

(ii) Generator operation

This condition is shown in Fig. 8.9(b). The position of m.m.f. phasors and the current distribution is obtained as follows:

(a) Since the machine supplies the power to the network and since the stator e.m.f. must always oppose the supply p.d. and cannot reverse, then the direction of the currents in the stator winding are reversed, i.e. the currents flow out of conductors R, Y and B and into conductors, R', Y' and B'. The m.m.f. phasor F_S is drawn from right to left.

(b) Since the rotor is driven at a higher speed than that of the magnetic field, its m.m.f. phasor F_R is now in front of F_S by $5\pi/6$ radians and *not* behind.

    The direction of the currents in the rotor conductors must therefore be as shown in Fig. 8.9(b). Note that the rotor revolves in an anticlockwise direction.

(c) Slip $s = - 0 \cdot 04$ and the frequency of rotor currents is $f_R = sf = - 0 \cdot 04 \times 50 = - 2$ Hz. therefore F_R rotates at 2 rev/s relative to the rotor itself, but in a *clockwise* direction.

    The speed of the rotor is $n_R = n_S - sn_S = 50 - (-0 \cdot 04) \, 50 = 52$ rev/s, and the speed of F_R with respect to the stator is now $52 - 2 = 50$ rev/s, i.e., the same as that of the stator m.m.f.

    Since F_R revolves clockwise with respect to the rotor, then the phase sequence in the rotor winding must be negative, i.e., the position of phases is interchanged. This is shown by marking the rotor conductors YY^1 and BB^1 which were marked BB^1 and YY^1 respectively during the motor operation.

(d) The air-gap m.m.f. F_A is still $62 \cdot 2$ A, but its position is now different. By applying Fleming's R.H. Rule to the rotor conductors, it is easily checked that the currents flowing in them will have the direction shown in Fig. 8.9(b). It must be remembered, however, that they move faster than the air-gap flux and therefore 'cut' it in an anticlockwise direction.

8.10   Operation of static induction machine

WINDING E.M.F.s

When a wound-rotor induction machine does not revolve and its rotor winding is not short-circuited, then the machine is termed an induction regulator.

    Under these conditions the slip is unity and no electro-mechanical power conversion takes place.

    The stator winding, connected to a three-phase a.c. supply, produces a sinusoidal rotating magnetic field as described in Chapter 6. The flux 'cuts' both the stator and the rotor conductors inducing in them sinusoidal e.m.f.'s.

    As the windings are distributed in slots around the circumference of the stator and the rotor, the e.m.f. equation 6.6 deduced in Chapter 6 can be applied as follows:

Stator e.m.f. in each phase

$$E_S = 4 \cdot 44 \, k_{s,S} \, k_{p,S} \, \Phi_A f N_S \text{ (V)} \tag{8.8}$$

and

Rotor e.m.f. in each phase

$$E_{R,o} = 4 \cdot 44 \, k_{s,R} \, k_{p,R} \, \Phi_A f N_R \text{ (V)} \tag{8.9}$$

Where E_S = r.m.s. value of winding e.m.f. induced in each stator phase

$E_{R,o}$ = r.m.s. value of winding e.m.f. induced in each rotor phase at standstill

$k_{s,S}$ = stator winding spread factor

$k_{s,R}$ = rotor winding spread factor

$k_{p,S}$ = stator winding pitch factor

$k_{p,R}$ = rotor winding pitch factor

Φ_A = air-gap flux per equivalent pole

f = supply frequency

N_S = number of turns per phase in stator winding

N_R = number of turns per phase in rotor winding

Dividing equation (8.9) by (8.8) gives

$$\frac{E_{R,o}}{E_S} = \frac{k_{s,R} k_{p,R} N_R}{k_{s,S} k_{p,S} N_S} = \text{constant} \tag{8.10}$$

Equation (8.10) indicates that the ratio between the rotor and the stator e.m.f.'s is constant at standstill and depends primarily on the turns ratio modified by spread and pitch factors of the windings.

The induction machine, therefore, behaves like a three-phase transformer. It must be noted that the factors for rotor and stator windings are not the same, because the number of slots in them may not be the same. It must also be remembered that when an induction transformer supplies a load current, the flux in the air-gap is due to the m.m.f. of the stator and the rotor currents.

SINGLE INDUCTION REGULATOR

A single induction regulator consists of a stator and a rotor each with a three-phase distributed winding. The stator winding is connected to a three-phase supply and the rotor winding gives a three-phase output of constant p.d. but of variable phase angle relative to the supply p.d. Flexible connections allow the rotor to be turned manually through an angle of π radians (180°) in either direction and thus phase differences between 0 and 2π radians are obtainable.

Figures 8.10(a) and (b) show the circuit and phasor diagram respectively for a basic regulator. An induction regulator in this form can only be used for phase-shifting purposes and therefore its use is not great. It finds application in laboratories, in a.c. meter testing and control of grids of mercury arc rectifiers.

The regulator, however, can also be made to produce a variable voltage output from a fixed voltage supply, when its windings are connected as in Fig. 8.11(a).

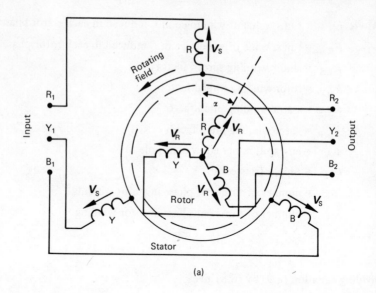

(a)

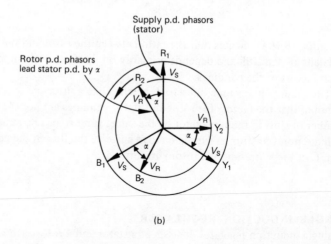

(b)

Fig. 8.10 Basic single induction regulator. (a) Circuit (b) Phasor diagram

This connection is termed a 'buck-boost' connection and is used to maintain the distribution network voltages within the prescribed limits.

The phasor diagram is given in Fig. 8.11(b). The phasor V_R rotates relative to the supply phasor V_S when the rotor is moved through $180°$. The voltage output therefore can be varied from a minimum equal to $V_S - V_R$ to a maximum equal to $V_S + V_R$. It must be noted that the phase angle between input and output p.d.'s also varies. To eliminate this a double induction regulator may be employed.

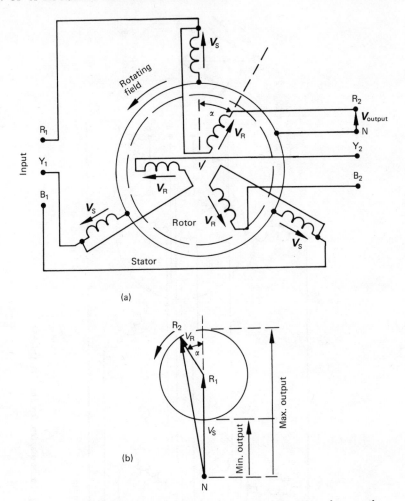

(a)

(b)

Fig. 8.11   (a) Circuit (b) Phasor diagram for phase R. 'Buck-boost' connection of an induction regulator

DOUBLE INDUCTION REGULATOR

This regulator is formed from linking electrically and mechanically two single induction regulators. The two rotors are assembled on a common shaft and the circuit diagram in Fig. 8.12(a) shows the interconnection of the windings.

The stator windings are both star connected but by interchanging the connections to Y and B phases of the second regulator, the direction of its rotating magnetic field is reversed. The phasor diagram in Fig. 8.12(b) shows the effect this has on the rotor phasors. The result is a constant-phase, variable voltage device with a reversible output with respect to the supply voltage phase position.

The functions of the stator and the rotor windings are often reversed in order to reduce the number of flexible connections to the rotor from six to three.

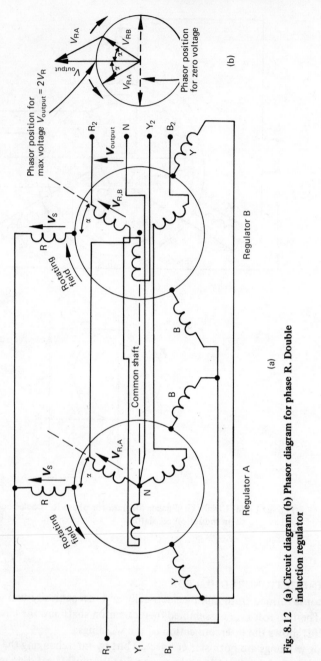

Fig. 8.12 (a) Circuit diagram (b) Phasor diagram for phase R. Double induction regulator

8.11 Operation of a rotating induction machine

When the rotor of an induction machine revolves the machine either converts electrical to mechanical energy (motor) or mechanical to electrical energy (generator).

In both cases the stator phase windings carry alternating currents and lie within the rotating magnetic field. Each conductor therefore experiences a force and has an alternating e.m.f. of rotation induced in it.

Thus the power conversion equation (3.14) of Chapter 3 can be applied to the air-gap region of an induction machine as follows:

$$3E_S I_S \cos \phi_S = T_S \omega_S \qquad (8.11)$$

where E_S is the stator e.m.f. per phase
$\quad\;\; I_S$ is the stator current per phase
$\quad\;\; \phi_S$ is the stator phase angle between E_S and I_S
$\quad\;\; T_S$ is the average torque of the rotating magnetic field,
and $\quad \omega_S$ is the speed of rotation of the magnetic field.

Factor of '3' is added because the machine has three windings.

The equations for e.m.f.'s, currents and the torque are deduced for both modes of operation below.

WINDING E.M.F.s AND ROTOR CURRENT

Whether the machine operates as a motor or a generator, the magnetic field rotates at the *same* speed which is determined by the frequency of the supply. Hence the expression for the *stator* e.m.f. is identical with that for an induction regulator and is given by the equation (8.8).

The frequency of the e.m.f.'s induced in the rotor winding however depends on the difference between the rotor's speed and the speed of the revolving field. It is given as $f_R = sf$ (equation (8.7)) and therefore the equation (8.9) must be modified accordingly:

Rotor e.m.f. in each phase

$$E_{R,s} = 4.44\, k_{s,R} k_{p,R} \Phi_A \, sf N_R = s E_{R,o} \;(\text{V}) \qquad (8.12)$$

where s = slip and E_{Ro} = rotor e.m.f. at standstill.

The voltage relationships for both windings of the machine can be considered now. They are obtained by applying Kirchhoff's Second Law to the induction machine's circuit in Fig. 8.13. Using the circuit marked for motor operation (Fig. 8.13(a)) the equations are as follows:

for the stator winding:

$$V_S = E_S + I_S R_S + I_S X_S \qquad (8.13)$$

for the short-circuited rotor winding:

$$E_{R,s} = I_{R,s} R_R + I_{R,s} X_{R,s} \qquad (8.14)$$

By reference to Fig. 8.13(b) the induction generator equations are:

for the stator winding:

$$V_S = E_S - I_S R_S - I_S X_S \qquad (8.15)$$

for the rotor winding:

$$E_{R,s} = I_{R,s} R_R + I_{R,s} X_{R,s} \qquad (8.16)$$

Examination of the circuit diagrams and equations shows that the stator current in the generator case has opposite reference direction to that of the motor, since the machine supplies the energy instead of drawing it from the supply. The equations (8.13) and (8.15) are, therefore, *not* the same.

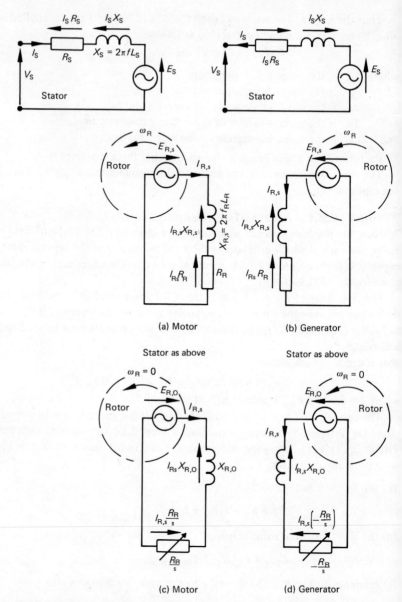

Fig. 8.13 **Circuit diagrams for rotating induction machine**

The equations for the rotor, however, are identical for the energy generated in its windings is dissipated as heat for both modes of operation.

All four equations represent phasor quantities and must be solved either graphically or by the usual methods of a.c. theory given in Chapter 2.

In order to obtain the torque equation, the expression for the rotor current is necessary and this is deduced by solving the equation (8.14) or (8.16) as follows:

Rotor current per phase

$$I_{R,s} = \frac{E_{R,s}}{(R_R^2 - X_{R,s}^2)^{\frac{1}{2}}} \text{ (A)}$$

Substituting $sE_{R,o}$ for $E_{R,s}$ (equation (8.12)) and $sX_{R,o}$ for $X_{R,s} = 2\pi sfL$, where $X_{R,o}$ is the standstill reactance of the rotor winding per phase,

$$I_{R,s} = \frac{sE_{R,o}}{(R_R^2 + s^2 X_{R,o}^2)^{\frac{1}{2}}} \text{ (A)}.$$

Dividing numerator and denominator by s, gives

$$I_{R,s} = \frac{E_{R,o}}{((R_R/s)^2 + X_{R,o}^2)^{\frac{1}{2}}} \quad\quad (8.17)$$

The equation (8.17) can be represented by modified circuits in Figs. 8.13(c) and 8.13(d). These show that:

(a) The rotor is now considered *stationary*, i.e., the rotating machine has been transformed into a static circuit, which nevertheless gives the same results as the original machine.

and (b) No electro-mechanical power conversion takes place in the machine. The power which would be converted into mechanical power in case of the motor or mechanical power which would be converted into electrical power in case of the generator, is now entirely dissipated in the new variable resistance (R_R/s). The new resistance is inversely proportional to the slip.
    It is to be noted that when the machine operates as a generator the slip is negative and the value of the resistance (R_R/s) becomes negative.

The negative resistance in this case can be interpreted as *producing* the energy rather than dissipating it.

MOTOR AND GENERATOR TORQUE

Using equation (8.11) the torque in rotating magnetic field is obtained as:

$$T_S = \frac{3}{\omega_S} E_S I_S \cos \phi_S \text{ (Nm)} \quad\quad (8.18)$$

It is usually expressed in terms of rotor parameters and the slip s. For this purpose the static circuit of the machine shown in Fig. 8.13(c) and (d) is used. It is clear that since the rotor is stationary, then the whole power in the rotating magnetic field must be equal to the electrical power dissipated in the variable resistance (R_R/s) of the three rotor windings, i.e.,

$$T_S = \frac{3}{\omega_S} I_{R,s}^2 \frac{R_R}{s}.$$

Substituting for $I_{R,s}$ from equation (8.17)

$$T_S = \frac{3}{\omega_S} \times \frac{E_{R,o}^2}{(R_R/s)^2 + X_{R,o}^2} \times \left(\frac{R_R}{s}\right)$$

i.e.

$$T_S = \frac{3}{\omega_S} \times \frac{E_{R,o}^2 \times R_R/s}{(R_R^2/s^2) + X_{R,o}^2} \quad (Nm) \tag{8.19}$$

It must be clearly understood, however, that the torque T_S of the rotating magnetic field *does not* give the value at which the electro-mechanical conversion takes place in a motor.

Let the conversion torque be denoted by T_R and let the speed of the rotor be ω_R, then the mechanical power at the shaft $T_R\omega_R$ is found to differ from $T_S\omega_S$ by the amount of power dissipated in the resistance of the short-circuited rotor winding, i.e.,

$$T_R\omega_R = T_S\omega_S - 3I_{R,s}^2 R_R \tag{8.20}$$

This occurs because the rotor winding in an induction machine *is not* supplied with electrical power by direct circuit connection but receives it from the rotating magnetic field via magnetic induction.

In case of a generator the conversion takes place at the value of torque given by T_S as seen in the next paragraph.

QUESTION 3. A three-phase, 50-Hz, 6-pole induction machine has the following particulars

Standstill rotor p.d. per phase	= 80 V
Rotor resistance per phase	= 0·4 Ω
Standstill rotor reactance per phase	= 12 Ω

(i) When the machine operates as a motor at 935 rev/min., calculate slip, rotor current and the torque transmitted from the magnetic field to the rotor.
(ii) When the machine is driven as an induction generator at 1100 rev/min., calculate the slip, the rotor current and the torque of the prime-mover. Neglect all losses.

ANSWER. *Motor operation*

$$\text{Speed of the magnetic field} = \frac{60f}{p} = \frac{60 \times 50}{3} = 1000 \text{ rev/min}$$

$$\text{Slip} = \frac{1000 - 935}{1000} = 0\cdot065 \text{ p.u.}$$

$$\text{Rotor current per phase} \quad I_R = \frac{80}{\sqrt{(0\cdot4/0\cdot065)^2 + 12^2}} = \frac{80}{\sqrt{6\cdot15^2 + 12^2}} = \frac{80}{13\cdot5}$$
$$= 5\cdot92 \text{ A}$$

$$\text{Torque} = \frac{3}{\omega_S} \times I_R^2 \frac{R_R}{s} = \frac{3}{(1000 \times 2\pi)/60} \times (5\cdot92)^2 \times 6\cdot15 = 6\cdot18 \text{ Nm.}$$

Generator operation

$$\text{Slip} = \frac{1000 - 1100}{1000} = -0\cdot1 \text{ p.u.}$$

$$\text{Rotor current per phase} \quad I_R = \frac{80}{\sqrt{(0\cdot4/0\cdot1)^2 + 12^2}} = 6\cdot31 \text{ A.}$$

$$\text{Torque} = \frac{3}{(1100 \times 2\pi)/60} \times (6\cdot31)^2 \times 4 = 4\cdot16 \text{ Nm.}$$

8.12   Power flow in an induction machine

An induction machine requires only three connections between the supply and the terminals of its stator winding.

The very simplicity of the electrical circuit makes the power flow through the machine more difficult to trace than in either direct current or synchronous machines. The power transfer for motor and generator operation is described separately:

INDUCTION MOTOR

When the slip is positive the machine operates as a motor and converts the electrical power to mechanical power.

The circuit diagram of one of its phases is shown in Fig. 8.13(a). The supply delivers to the machine three times the power absorbed by one phase, i.e.

$$\text{electrical power input} = 3V_S I_S \cos \phi_i \text{ (W)}$$

where $\cos \phi_i$ is the input power factor.

Not all of this is transformed into power in the rotating magnetic field, because a portion is lost in heating:

    (i) the three stator winding resistances, i.e., stator resistive loss $= 3I_S^2 R_S$

and (ii) the stator core loss, due to eddy currents and hysteresis i.e. P_S

Thus the power transmitted by the rotating field is:

$$T_S \omega_S = 3V_S I_S \cos \phi_i - 3I_S^2 R_S - P_S \tag{8.21}$$

The power $T_S \omega_S$ is then converted into

    (i) the mechanical power of the rotor $T_R \omega_R$, which is reduced by the rotational loss C due to friction in bearings and in air.

and (ii) the electrical power made up of a resistive loss $3I_{R,s}^2 R_R$, which is dissipated in heating the resistance of the rotor winding, and a very small rotor core loss P_R (that is eddy current and hysteresis loss).

If T_M is the output torque obtainable from the rotor shaft, then the power equation for the rotor is

$$T_M \omega_R = T_S \omega_S - 3I_{R,s}^2 R_R - P_R - C \tag{8.22}$$

The block diagram in Fig. 8.14 shows clearly the power flow and the conversion from electrical to mechanical form via the magnetic link.

INDUCTION GENERATOR

When slip is negative, the rotor of an induction machine is driven at a speed ω_R greater than the speed ω_S of its magnetic field. This is achieved by a prime-mover supplying a mechanical power to the machine, which in turn converts it into electrical power and feeds it into a distribution network. The machine is termed an induction generator. Figure 8.13(b) shows one phase of the machine marked for generator operation.

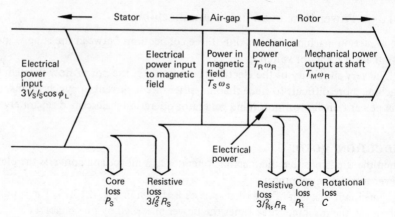

Fig. 8.14 Power flow through the induction motor

Since the rotor conductors move faster than the magnetic field, the direction of magnetic flux 'cutting' is reversed and in consequence rotor e.m.f. and currents $I_{R,s}$ and I_S are all reversed. It is now clear that since the arrows of E_S and I_S agree in direction, the stator winding is a source of electrical power (see Chapter 1). The arrows of $E_{R,s}$ and $I_{R,s}$ also agree, because the electrical power generated in the rotor is lost in heating its winding. The prime-mover must now overcome the magnetic forces due to currents in the stator and rotor windings and its rotation is anticlockwise, i.e., the same as in the motor case.

The power flow diagram for an induction generator is shown in Fig. 8.15. The diagram clearly indicates that the machine receives electrical volt-amperes from the supply to produce the rotating magnetic field. The power required to establish the flux is, however, zero since the exciting current is $\pi/2$ (rad) out of phase with the supply voltage.

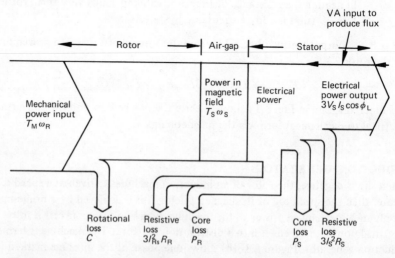

Fig. 8.15 Power flow through the induction generator

The power equations are as follows:

$$T_S \omega_S = T_M \omega_R - C \qquad (8.23)$$

where T_M is the mechanical torque applied to the generator shaft by the prime-mover, and

$$3 V_S I_S \cos \phi_i = T_S \omega_S - 3 I_S^2 R_S - P_S - 3 I_{R,s}^2 R_R - P_R \qquad (8.24)$$

where $3 V_S I_S \cos \phi_i$ is the electrical output of the generator.

8.13   Losses in induction machines

In the previous paragraph all the losses occurring in the induction machine were identified. They may be grouped under two headings:

   (a) Fixed losses, which consist of
     (1) Stator core loss P_S
     (2) Friction and windage C
and (b) Variable losses, which comprise
     (1) Stator resistive loss $3 R_S I_S^2$
     (2) Rotor resistive loss $3 R_R I_{R,s}^2$
     (3) Stray losses which include the core loss P_R in the rotor.

The induction machine normally operates from a supply of constant voltage and frequency. Since the stator core loss (hysteresis and eddy current) is dependant on frequency and maximum flux density, it remains unchanged under different load conditions, because the rotating magnetic flux is proportional to the supply voltage (equation (8.8)).

Furthermore, the speed of rotation varies very little between no load and full load as seen from typical slip values given in paragraph 8.6. Hence friction and windage loss is almost constant, since it depends on the speed of rotation.

In contrast, losses b(i) and b(2) vary over a very wide range, since the stator and rotor currents are different for each value of mechanical or electrical loading. The stray loss b(3) although variable, is very small and is usually neglected. This is justified since on load (i) the frequency of the currents in the rotor winding is very low and (ii) stray flux inducing eddy currents in the stator frame and endshields is not of appreciable magnitude.

The fixed losses are obtained from a no load test. The machine is run unloaded from the supply at rated voltage and frequency and the power input to it is measured. Since the machine does not deliver power, the stator and rotor currents are very small and therefore the $I_S^2 R_S$ and $I_{R,s}^2 R_R$ loss may be neglected and the whole power input is dissipated in losses a(1) and a(2). The stator resistive loss can be calculated for various values of stator current I_S, measured by an ammeter connected in a 'line to stator' lead. The resistance R_S is obtained by any of the normal methods of measurement such as Ammeter/Voltmeter or Ohmmeter. R_S however should be measured hot, after the machine has been running for some time at full load.

If the rotor has a phase winding brought out to sliprings then the rotor resistive loss can be similarly estimated. In the case of the cage type rotor the following is used:
From paragraph 8.11,

$$3I_{R,s}^2 \frac{R_R}{s} = T_S \omega_S$$

therefore $$3I_{R,s}^2 R_R = sT_S \omega_S \tag{8.25}$$

i.e., rotor resistive loss is equal to slip multiplied by the power in the magnetic field $T_S \omega_S$.

8.14   Efficiency of an induction machine

The per unit efficiency can be calculated from the equations (1.6) and (1.7) given in Chapter 1. The mechanical input or output at the shaft is taken as $T_M \omega_R$ (Nm/s)

where T_M = shaft torque in Nm.
and ω_R = speed of rotation in rad/s.

The electrical input or output at the stator terminals is

$$3V_S I_S \cos \phi_i = \sqrt{3} V_L I_L \cos \phi_i \text{ (W)}$$

where V_L = Line voltage in volts
                   I_L = Line current in amperes
and $\cos \phi_i$ = power factor.

When the machine operates as a motor the power factor is always lagging and when it operates as an induction generator it is always leading. The reason for the leading power factor in the latter case is that the machine *does not* provide its own excitation, but takes it from the supply to which it is connected.
Thus:

$$\text{Motor efficiency} = \frac{T_M \omega_R}{\sqrt{3} V_L I_L \cos \phi_i} \text{ p.u.} \tag{8.26}$$

and

$$\text{Generator efficiency} = \frac{\sqrt{3} V_L I_L \cos \phi_i}{T_M \omega_R} \text{ p.u.} \tag{8.27}$$

QUESTION 3. A 40-kW, 8-pole induction machine is connected to a 415-V, 3-phase, 50 Hz supply. When working as a motor on full load its efficiency is 0·85 p.u., its power factor 0·8 lagging and its slip 0·03.
    When working as a generator at a slip of $-0·04$, its power factor is 0·78 leading and its efficiency 0·82 p.u.
    Find for each mode of operation:

   (i)   rotor speed in rev/min
   (ii) rotor frequency
   (iii) torque at the shaft
   (iv) line and phase currents assuming the stator to be connected in delta.

ANSWER. *Motor operation*

speed of magnetic field $N_S = \dfrac{60 \times 50}{4} = \underline{750 \text{ rev/min}}$.

(i) rotor speed $N_R = 750 - 0.03 \times 750 = \underline{722.5 \text{ rev/min}}$

(ii) rotor frequency $f_R = 0.03 \times 50 = \underline{1.5 \text{ Hz}}$.

(iii) shaft Torque $T_M = \dfrac{40\,000}{\omega_R} = \dfrac{40\,000}{722.5 \times 2\pi/60} = \underline{530 \text{ Nm}}$

input power $= \dfrac{40\,000}{0.85} = \underline{47\,100 \text{ W}}$

(iv) line current $I_L = \dfrac{47\,100}{\sqrt{3} \times 415 \times 0.8} = \underline{82 \text{ A}}$

phase current $I_{ph} = \dfrac{82}{\sqrt{3}} = \underline{47.1 \text{ A}}$

Generator operation

(i) rotor speed $N_R = 750 + 0.04 \times 750 = \underline{780 \text{ rev/min}}$

(ii) rotor frequency $f_R = 0.04 \times 50 = \underline{2 \text{ Hz}}$

(iii) shaft torque $T_M = \dfrac{40\,000}{0.82 \times 780 \times 2\pi/60} = \underline{600 \text{ Nm}}$

output power $= \underline{40\,000 \text{ W}}$.

(iv) line current $I_L = \dfrac{40\,000}{\sqrt{3} \times 415 \times 0.78} = \underline{71.6 \text{ A}}$

and phase current $I_{ph} = \dfrac{71.6}{\sqrt{3}} = \underline{41.3 \text{ A}}$

In the above example it is assumed that the machine is fully loaded as a generator when supplying 40 kW to the supply network.

QUESTION 4. Calculate the full load efficiency of an induction motor from the following information:

Line Voltage:	415 V	F.L. Line current:	50 A
Number of phases:	3	F.L. power factor:	0.85
Frequency:	50 Hz	F.L. slip:	0.04

Stator winding: delta connected and its resistance = $0.52 \ \Omega$ per phase.

When the no-load test was performed on the machine, the results obtained were as follows:

Power intake = 1550 W

and the ratio of $\dfrac{\text{stator core loss}}{\text{friction and windage loss}} = \dfrac{3}{2}$

ANSWER.   F.L. input to motor $= \sqrt{3} \times 415 \times 50 \times 0.85 = 30\,000 \text{ W}$

F.L. stator resistive loss $= 3 \times 0.52 \left(\dfrac{50}{\sqrt{3}}\right)^2 = 1300 \text{ W}$

stator core loss $= \frac{3}{5} \times 1550 = 930$ W

Stator losses $= \underline{2230$ W$}$

power transferred to rotor via magnetic field $= \underline{27\ 770$ W$}$

rotor resistive loss $= 0.04 \times 27\ 770 = 1110 \cdot 8$ W

friction and windage loss $= \frac{2}{5} \times 1550 = 620$ W

rotor losses $= \underline{1730 \cdot 8$ W$}$

mechanical output at shaft $= \underline{26\ 040 \cdot 2$ W$}$

efficiency $= \dfrac{26\ 040 \cdot 2}{30\ 000} = \underline{0 \cdot 868 \text{ p.u.}}$

8.15   Characteristics of induction machines

The induction machine whether operating as a motor or a generator must *always* be connected to a supply network of which synchronous machines (discussed in the next chapter) *must* also be a part. The network voltage and frequency are therefore *constant* and are not changed by the behaviour of an induction machine.

In these circumstances the operating characteristic of the induction generator cannot be expressed by a graph of its terminal p.d. plotted against the load current as for d.c. generators, for the machine's terminal voltage is constant and always equal to that of the supply.

From previous paragraphs it is known that for the machine to generate electrical power it must be driven by a mechanical prime-mover at a speed greater than that at which the magnetic field rotates. Hence it is the torque of the prime-mover and the speed at which it drives the induction generator that influences its operation. When the same machine operates as a motor its characteristic is the curve relating output torque at shaft to the speed of rotation, just as in the case of a d.c. motor. Thus the 'load characteristic' for an induction machine, whether generating or motoring is given by one graph of the rotor speed ω_R plotted against the shaft torque T_M. Such a graph obtained experimentally is shown in Fig. 8.16(a), and the corresponding curve of the load current flowing in one phase of the stator winding is plotted in Fig. 8.16(b).

When the three-phase supply is switched on to the unloaded machine, the starting torque OP is instantly generated and accelerates the rotor along the curve until point 'S'. At 'S' the machine produces torque SS' which equals the torque due to opposing losses, and the motor rotates at constant speed. The speed is nearly equal to that of the magnetic field. When the mechanical load is now applied to the shaft, the motor must deliver a larger torque. If the load torque is equal to TT', then the speed of the motor drops slightly until point T and again the motor operates stably at constant speed. The load torque can be increased further up to a maximum equal to QQ', which the machine is capable of providing.

If this value is exceeded the machine will stall and eventually stop. Now let the prime-mover apply a driving torque to the shaft of the machine equal to UU'. This torque must now be opposed by the torque generated by the machine itself which

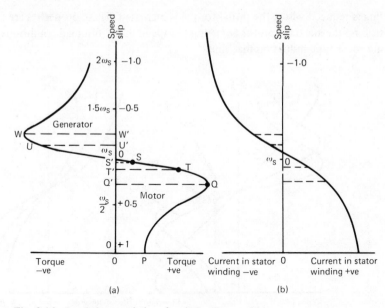

Fig. 8.16 Load characteristics of an induction machine

therefore becomes negative. Its values are therefore plotted to the left of the speed axis. The speed of the rotor rises above ω_S and the mechanical power equal to $T_{uu}\omega_u$ (W) is converted to the electrical power $3V_SI_S \cos \phi_i$, plus losses. The limit at which the machine can be operated is indicated by WW′. Any further increase in the driving torque will meet with a smaller opposing torque produced by the induction machine, and results in a smaller power being supplied to the network. The rotor speed will thus continue to rise. Hence the working range of the induction generator is given by that portion of the curve between points W and Q.

In Fig. 8.6(b) the values of phase current flowing in the stator winding are shown. At starting the current taken from the supply is a maximum. Then it decreases as the speed of the rotor increases until zero is reached at a speed just above ω_s. At this point the prime-mover supplies all the losses to the machine. Further increase in speed results in a changed direction of power flow, with the supply system now receiving the energy. The current is therefore plotted to the left of the speed axis. The current magnitude now increases with speed.

Reference to equations (8.17) and (8.19) indicates that the rotor current and therefore stator current as well as the torque depend on the value of R_R/s.

If all factors in these equations are unchanged except R_R then automatically the value of slip s will adjust itself so that R_R/s is still the same. The influence of changing the value of rotor winding resistance R_R on the speed/torque and speed/stator current characteristics is indicated in Fig. 8.17(a) and (b).

It is obvious from these that no new values of current or torque can be obtained, but the same magnitudes occur at different speeds. For instance, the maximum torque is still equal to QQ′ but with the resistance value being larger it now occurs at a much lower speed (bigger slip) as seen from curve C. The speed range at which the machine now operates stably is greatly increased. Furthermore the stator current at starting is reduced whilst the initial torque is increased. These properties are used for controlling the starting and for adjusting the speed under full load conditions, in a wound rotor type induction machine.

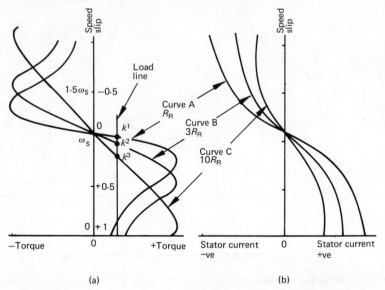

Fig. 8.17   Influence of rotor winding resistance on the characteristics of induction machines

In cage rotor machines the profile of the slots of the cage winding is determined at the design stage to obtain the desired operating characteristics. The three main types of slot design, designated A, B and C, are shown in Fig. 8.18(a), whereas the characteristics obtainable with each are drawn in Fig. 8.18(b).

At starting and at high values of slip the magnetic flux concentrates near the top of the rotor slot so that the action is limited to this section of the conductor. At normal running speeds the whole conductor is active. The cage winding with cross-section A has a relatively low resistance and gives the machine with the usual characteristic. Cage B rotor has a high resistance at starting and low resistance during running giving a machine with high starting torque and a low initial current. Cage C machine has a high rotor resistance at all times and is used for high slip application.

Slot design B maybe thought of as consisting of two separate cage windings; one outer which operates at starting and the other inner which is active during normal running. An induction machine with two separate cages however is rarely built nowadays since the improved design of single cage rotors have practically eliminated its use.

The shape and size of the conductor section and the use of different materials enables the designer to choose the right combination to provide the characteristics necessary to meet the required duty.

8.16   Control of induction machines – generators

As with d.c. generators the control of an induction generator involves the ability to determine the amount of electrical power to be supplied to the network.

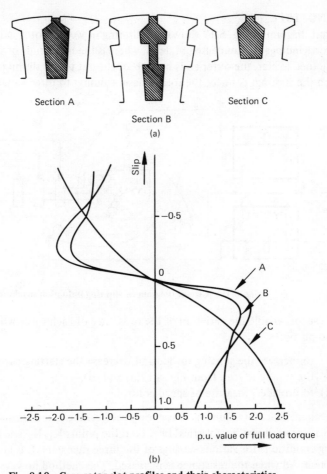

Fig. 8.18 Cage rotor slot profiles and their characteristics

The machine may be started as an induciton motor, before the mechanical prime-mover develops a driving torque, or it may be driven to near the synchronous speed ω_s and then connected to the supply terminals. No synchronisation is necessary and the only precaution to be taken is that the prime-mover and the magnetic field in the machine's gap rotate in the *same* direction. Since the amount of electrical power supplied to the network by the induction generator depends on the torque and the driving speed of the prime-mover, the control is wholly confined to the mechanical side. As such therefore, it lies within the sphere of mechanical engineering depending as it does on the type of the mechanical device used, such as a petrol or diesel engine, gas turbine, water turbine etc.

8.17   Control of induction machines – motors

The control involves the ability to start, stop and vary the speed of these types of induction motors.

SLIP-RING TYPE

The circuit diagram in Fig. 8.19 shows the slip-ring or wound rotor induction machine. The stator windings are connected in delta, whereas the inner ends of the rotor windings are joined in star, the outer ends being brought out to the sliprings. The brushes resting on the sliprings connect the rotor phase windings to three variable resistors.

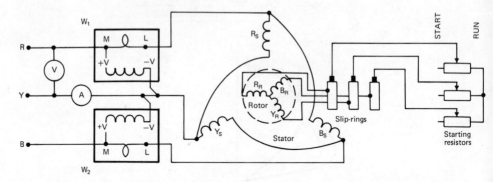

Fig. 8.19 **Circuit diagram of slip-ring induction machine**

The resistors enable the variation of the resistance of each rotor winding R_R to serve two purposes:

     (i) to increase the starting torque and decrease the starting current taken from the supply as seen from Fig. 8.17(a) and (b).

and   (ii) to provide a measure of speed control.

If the mechanical load torque line is superimposed on the speed/torque characteristics in Fig. 8.17(a) as shown by a vertical line, then the points K_1, K_2 and K_3 give the operating conditions for various settings of the three rheostats. It is clear that the speed is reduced when the resistance is increased. This method however, is wasteful of energy, which is lost in heating the rheostats, and therefore reduces the amount converted to useful mechanical work.

The instruments shown in the circuit read the following quantities:

Voltmeter – Line voltage,
Ammeter – Line current,

and the sum of the two wattmeter indications = total electrical power input.

The speed/torque characteristics in Fig. 8.17(a) can be obtained by loading the machine with a brake and measuring the output torque and the speed of rotation.

QUESTION 5. A 7·5 kW slip-ring induction motor gave the following results when tested under various load conditions:

    (i) Starting resistors short-circuited.

Torque in (Nm)	80	100	125	150	175	150	125	100	75	50	25	0
Speed in (rev/min)	0	420	730	910	1160	1300	1350	1380	1420	1450	1470	1500

(ii) Full Starting resistors in rotor circuit.

Torque in (Nm)	150	170	175	170	150	125	100	75	50	25	0
Speed in (rev/min)	0	170	290	400	600	760	920	1060	1200	1340	1500

Plot the speed/torque characteristics from the above results and estimate for each case.

(1) starting torque
(2) maximum torque and the power output at this torque.
(3) torque, speed and slip to give rated output.

ANSWER

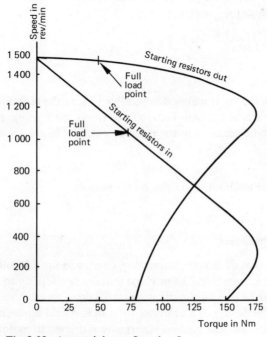

Fig. 8.20 Appertaining to Question 5

(i) *Starting resistors out – from the graph*
   1. starting torque = <u>80 Nm.</u>

   2. maximum torque = <u>175 Nm</u> and occurs at 1160 rev/min.

   $$\therefore \text{ maximum power output} = 175 \times \frac{1160 \times 2\pi}{60} = \underline{21 \cdot 1 \text{ kW}}$$

   3. $T_M \omega_R = 7500$ W.

   $$\therefore T_M \times \frac{2\pi N_R}{60} = 7500 \text{ W.}$$

   and $T_M N_R = 71700$

A point on the graph is chosen, i.e. when $T_M = 60$ Nm and $N_R = 1440$ rev/min. $T_M N_R = 60 \times 1440 = 86\ 500$ which is too high. Trying again: $T_M = 49 \cdot 5$ Nm and $N_R = 1450$ rev/min hence this is the full load operating point.

The slip $s = \dfrac{1500 - 1450}{1500} = 0 \cdot 033$ p.u.

(ii) *Starting resistors in – from the graph*
    1. starting torque = 150 Nm.
    2. maximum torque = 175 Nm at 290 rev/min.

$$\therefore \text{ maximum power output} = 175 \times \frac{290 \times 2\pi}{60} = 5 \cdot 33 \text{ kW}$$

    3. $T_M N_R = 71\ 700$ hence by estimate from the graph
       $T_M = 65$ Nm and $N_R = 1100$ rev/min.

$$\therefore \text{ slips} = \frac{1500 - 1100}{1500} = 0 \cdot 267 \text{ p.u.}$$

CAGE TYPE

The construction of the rotor makes it impossible to alter the resistance of the cage winding once the machine has been built, and therefore the starting, stopping and speed control has to be confined to the machine's input side.

STARTING

There are three basic methods of starting a cage motor:

       (i) direct on line
      (ii) star-delta
and (iii) auto-transformer.

The first method is that of directly connecting a motor to the full voltage of the supply by means of a contactor. The machine initially develops up to twice its full load torque and takes a very large current, up to 10 times its rated value. The current quickly diminishes as the motor accelerates to its near synchronous speed. Because of the initial current 'rush' the method is normally restricted to motors below 4 kW rating, fed from a consumer network, to avoid supply voltage fluctuations and its consequent harmful effects on other plant. In electric power stations and in heavy industries which are supplied at High Voltage, much larger machines may be started in this way, their rating being of the order of hundreds of kW's.

    Star-delta starting is used for machines rated above approximately 4 kW, in order to limit the initial current. In Fig. 8.21 a circuit diagram shows the connection of the motor with a star-delta starter. In this method the stator windings are connected in 'STAR' for STARTING and then when the machine is running, quickly reconnected in DELTA for normal operation.

    From theory in Chapter 2, in the START position each stator winding has $1/\sqrt{3}$ of its normal voltage applied to it and therefore the line current is

$$\frac{1}{\sqrt{3}} \times \frac{1}{\sqrt{3}} = \frac{1}{3}$$

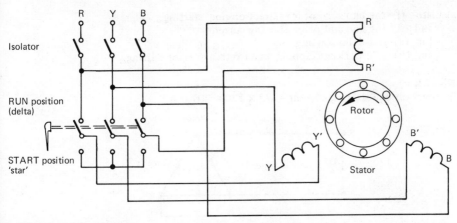

Fig. 8.21   Star-delta starting

of the value in delta. The starting torque is also $\frac{1}{3}$ of its delta direct on line value. The reduction in torque is of course a disadvantage where quick run up to full speed is required but has to be accepted where current limitation is necessary.

Auto-transformer starting, also reduces the initial voltage applied to the motor and therefore the torque and the initial current taken from the supply. The advantage of the method is that current reduction can be adjusted to the required value by choosing the correct tapping on the auto-transformer, unlike the previous method where the current reduction is fixed at $\frac{1}{3}$ of its normal value. Figure 8.22 shows the machine with an auto-transformer starter.

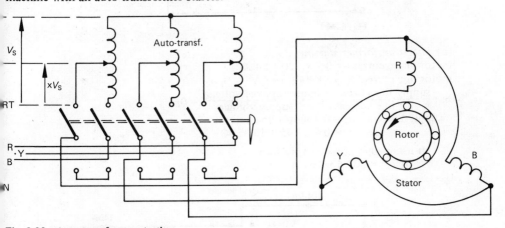

Fig. 8.22   Auto-transformer starting

If the fractional tapping is 'x', usually 0·7 to 0·8 of the full load voltage, then the starting torque is reduced to x^2 times its direct on line value and the starting current is approximately x^2 times its line value, as seen from the following example:

QUESTION 6. A three-phase delta-connected cage-type induction motor when connected directly to a 415-V, 50-Hz supply takes a starting current of 120A in each stator phase.

Calculate   (i) the line current for 'direct-on-line' starting,
          and (ii) the line and phase starting currents
          for (a) star-delta starting,
          and (b) a 70 per cent tapping on auto-transformer starting.

ANSWER
(i) direct on line starting current = $\sqrt{3} \times 120 = \underline{207 \cdot 6 \text{ A}}$

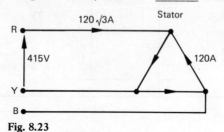

Fig. 8.23

(ii) (*a*) *Star-delta starting.*
      Since 415 V produce 120 A in phase winding,
      hence $415/\sqrt{3}$ produces $120/\sqrt{3} = \underline{69 \text{ A}}$

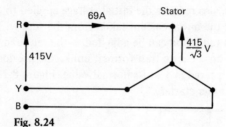

Fig. 8.24

(*b*) *auto-transformer starting*
motor phase current = $0 \cdot 7 \times 120 = 84$ A
motor line current = $\sqrt{3} \times 84 = 145$ A
but $\dfrac{\text{supply line current}}{\text{motor line current}} = \dfrac{\text{motor applied voltage}}{\text{supply voltage}} = 0 \cdot 7$
∴ supply line current = $0 \cdot 7$ motor line current
                  $= 0 \cdot 7 \times 145 = \underline{101 \cdot 5 \text{ A}}.$

   or using approximate expression
      $(0 \cdot 7)^2 \times 207 \cdot 5 = \underline{101 \text{ A}}$

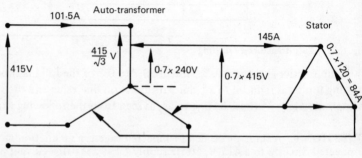

Fig. 8.25

SPEED CONTROL

Combining equations (8.4) and (8.5) the speed of an induction machine can be expressed as:

$$n_R = (1 - s)\frac{f}{p} \text{ (rev/s)} \qquad (8.6)$$

The equation shows that the speed depends on

    (a) slip s
    (b) number of equivalent pole-pairs p
and   (c) supply frequency f

The ability to vary anyone of the above three quantities will provide the means of speed control.

The slip can be changed by adjusting the resistance of the rotor winding and this method is employed in wound rotor induction machines as already described. In cage type machines this cannot be done, and therefore variations of either p or f are the only alternatives.

If two or three operating speeds are required near a possible synchronous speed, pole changing can be used. There are three main ways in which this can be done. The first is to wind the stator of a cage induction machine with two separate windings, each designed for a different speed. The windings are brought out to external terminals and one or the other can be selected by means of manual or motor-driven selector switch. With this method two preselected speeds can be obtained. For instance 2, 4, 6 or 8 equivalent pole windings will give speeds of magnetic field of 3000, 1500, 1000 or 750 rev/min, at a supply frequency of 50 Hz.

The second method consists of dividing a single stator winding into two portions which can be connected to give either p or $2p$ pairs of poles. This arrangement enables the two speeds in the ratio of 2:1 to be obtained. Machines which combine the two methods are also constructed, and will operate at three or four fixed speeds.

The third method of pole changing developed in recent years is termed *pole amplitude modulation*. The method is based on subdividing the stator winding into conductor groups which can be connected externally in different ways to give speed ratios such as 6:8, 8:10, 10:12 etc. As such this is not strictly a pole-changing method for the speeds which can be achieved are not restricted to those determined by an even number of poles. So far pole amplitude modulation has been used in large induction machines.

The most efficient way of controlling the speed of an induction machine is by means of a variable frequency supply. The method requires some form of frequency convertor together with a means of adjusting the voltage. The latter is necessary because the reactance of the machine varies with frequency. As the frequency is increased the supply voltage must also increase; when the frequency is lowered the voltage must also be lowered in proportion.

At the present time the variable frequency is usually obtained from solid-state equipment, although rotary converters (motor-generator sets) are still used occasionally.

Equipment employing silicon rectifiers converts the normal three-phase 50-Hz supply to direct current and then inverts it using thyristors to produce a.c. at the required frequency. With this method any desired frequency may be obtained either below or above the speed of the magnetic field for which the machine has been constructed. The alternative method of frequency control consists of using thyristors and triacs to cut the original 50 Hz sinewave into segments and to sum the segments again into a roughly defined sinewave of lower frequency. With this approach only speeds below that of the machine's field can be obtained.

Both methods which utilise solid state devices, produce jagged voltage wave shapes, and cause increased losses in machines. The machines which are to operate from variable frequency supplies must therefore be either designed for this duty or normal machines should be derated.

STOPPING

An induction motor, whether wound-rotor or cage type is stopped normally by disconnecting it from the supply. If, however, it is important to stop it very quickly, then the phase sequence of the supply to the stator is reversed by interchanging any two supply connections. This reverses the direction of rotation of the magnetic field and develops a braking torque on the rotor. The method is called plugging.

8.18   Applications

INDUCTION MOTORS

The majority of induction machines employed today are operated as motors and by comparison their use as induction regulators and even more so as induction generators is insignificant.

As a nearly constant speed motor, an induction machine has no competitor. Other types of machines are considered only, when a variable speed drive is required

The applications of three-phase induction motors, the size of which range from half a kilowatt to tens of thousands of kilowatts, are therefore so numerous that they are found everywhere from a smallest workshop to the largest factory in steel, manufacturing and service industries, gas, water and electricity undertakings, road and sea transport etc.

In fact, wherever an electrical drive is needed the induction motor is the first to be considered and generally fullfils the requirements.

INDUCTION GENERATORS

Induction machines are very rarely built to operate solely as generators. Although their operation in this mode is very simple, their control being confined to the mechanical prime-mover, nevertheless the inability to vary the voltage and the power factor (always leading) of the machine militates against their use.

Because of this, the generation of electrical power is limited to schemes where an induction machine is operated as a motor and as a generator alternately. Some examples of these are:

(i) mountain trains, where during ascent the induction motors produce motive power and during the downward journey they operate as generators driven by the weight of the train and thus feed electrical power into the network and also act as a brake.

    (ii) crane drives where similar sequence of operation of induction machine occurs during lifting and lowering of the loads.

   (iii) two-level water reservoir electrical power stations. In these schemes the machines operating as motors driving water pumps, lift the water from lower to higher level reservoirs during electrical off peak periods. During periods of high demand for power, the water is released from higher level reservoirs and drives the pumps or turbines which in turn rotate the induction machines at above synchronous speed, thus feeding electrical energy into the network.

      Similar schemes may be used in ocean tidal generating stations.

INDUCTION REGULATORS

Many induction regulators are used for direct control of the voltage of a given system. In this application illustrated by a 'buck-boost' connection shown in Fig. 8.11, the regulator is inserted between the supply and the load and influences the output voltage directly. Many variants of this scheme are employed according to different requirements and conditions. However, by far the greatest number of induction regulators is employed indirectly in controlling a required quantity. Typical examples are where the regulators control:

    (i) the supply voltage to static capacitors and thereby adjust the *power factor* of an installation,

   (ii) the supply voltage to heaters and thus adjust the room *temperature*,

  (iii) the *speed* of rotation of a.c. commutator motors and the *frequency* of rotary frequency convertors,

  (iv) the phase shift of the applied voltage to the grids of mercury arc rectifiers and therefore control their *d.c. output*,

   (v) the excitation of synchronous machines using static rectifiers and thereby their *voltage, power* and *power factor* (see Fig. 9.22).

    At present the largest application of indirect control by means of induction regulators is still in the speed control of a.c. commutator machines.

SUMMARY

1. An induction machine is defined as having a phase winding on the STATOR and a phase or cage winding on the ROTOR. All windings carry alternating currents. There are therefore two basic types of induction machine

    (i) Squirrel cage machine (cage type rotor)
   (ii) Slip-ring machine (phase wound rotor)

2. Wound rotor induction machines can be operated in three different modes:

      (i) as a motor
     (ii) as a generator
and (iii) as an induction regulator.

A cage rotor machine can only operate as (i) or (ii).

3. The speed of rotation of the magnetic field is given by:

$$\omega_S = \frac{2\pi f}{P} \text{ (rad/s)} \tag{8.3}$$

or $\qquad n_S = \dfrac{f}{p}$ (rev/s) $\hfill$ (8.4)

where $\quad f$ = supply frequency in Hz,

and $\qquad p$ = number of equivalent pole pairs.

4. Slip-speed = $(\omega_S - \omega_R)$ rad/s = $(n_S - n_R)$ rev/s

and Slip $s = \dfrac{\omega_S - \omega_R}{\omega_S} = \dfrac{n_S - n_R}{n_R}$ p.u. $\hfill$ (8.5)

where ω_S = Synchronous speed of the magnetic field in rad/s

$\qquad n_S$ = Synchronous speed of the magnetic field in rev/s

$\qquad \omega_R$ = rotor speed in rad/s

$\qquad n_R$ = rotor speed in rev/s

$\qquad s$ is a ratio and has no units

when s is positive, the machine operates as a motor

when s is negative, the machine operates as a generator

when s is unity, the machine operates as induction regulator.

5. $\qquad$ Stator frequency = supply frequency = $f = \dfrac{p\omega_S}{2\pi}$ (Hz) $\hfill$ (8.6)

$\qquad$ Rotor frequency = sf (Hz) $\hfill$ (8.7)

6. $\qquad$ Stator e.m.f. per phase $E_S = 4.44\, k_{s,S} k_{p,S} \Phi_A f N_S$ (V) $\hfill$ (8.8)

$\qquad$ Rotor e.m.f. per phase $E_{R,s} = 4.44\, k_{s,R} k_{p,R} \Phi_A sf N_R$ $\hfill$ (8.12)

$\qquad\qquad\qquad = sE_{R,o}$ (V)

where E_S $\quad$ stator e.m.f. per phase in V

$\qquad E_{R,s}$ = rotor e.m.f. per phase in V

$\qquad E_{R,o}$ = rotor e.m.f. per phase at standstill in V

$\qquad k_{s,S}$ = stator winding spread factor

$\qquad k_{s,R}$ = rotor winding spread factor

$\qquad k_{p,S}$ = stator winding pitch factor

$\qquad k_{p,R}$ = rotor winding pitch factor

$\qquad \Phi_A$ = air-gap flux per equivalent pole in Wb

$\qquad f$ = supply frequency in Hz

$\qquad N_S$ = number of turns per stator phase winding

$\qquad N_R$ = number of turns per rotor phase winding

$\qquad s$ = slip

7. The stationary induction machine is termed on induction regulator. A single induction regulator can be used to produce an adjustable phase shift between the supply and the load or as a variable voltage device when a 'buck-boost' connection is used.

Two induction regulators connected mechanically and electrically give a variable voltage device without introducing phase shift.

8. $\qquad$ Rotor current per phase $I_{R,s} = \dfrac{E_{R,o}}{((R_R/s)^2 + X_{R,o}^2)^{\frac{1}{2}}}$ $\hfill$ (8.17)

where R_R $\quad$ = rotor winding resistance per phase in ohms

$\qquad X_{R,o}$ = rotor winding leakage reactance per phase at standstill in ohms.

9. Torque exerted by the rotating magnetic field on the rotor is obtained from

$$E_S I_S \cos \phi_S = T_S \omega_S \qquad (8.11)$$

and is given by:

$$T_S = \frac{3}{\omega_S} E_S I_S \cos \phi_S = \frac{3}{\omega_S} \cdot \frac{E_{R,o}^2 R_R/s}{R_R^2/s^2 + X_{R,o}^2} \text{ (Nm)} \qquad (8.18/8.19)$$

The torque at which electro-mechanical power conversion takes place is T_R for the motor case and is obtained from

$$T_R \omega_R = T_S \omega_S - I_{R,s}^2 R_R \qquad (8.20)$$

For the generator the conversion takes place at the value of torque given by T_S.

10. Efficiency of an induction motor $= \dfrac{T_M \omega_R}{\sqrt{3} \, V_L I_L \cos \phi_i} \text{ p.u.}$ \qquad (8.26)

  Efficiency of an induction generator $= \dfrac{\sqrt{3} \, V_L I_L \cos \phi_i}{T_M \omega_R} \text{ p.u.}$ \qquad (8.27)

where T_M = shaft torque in Nm
                V_L = line voltage in V
                I_L = line current in A
and $\cos \phi_i$ = phase power factor.

11. Characteristics are graphs which relate torque T_M to the speed of the rotor ω_R for the induction machine working as a motor *and* as a generator, since in the latter case the output voltage is constant and determined by the supply to which the induction generator is connected. An induction generator cannot normally operate in isolation, that is, supply a passive load because it does not have its own excitation and must draw it from a three-phase supply network.

12. The induction generator is controlled by adjusting the mechanical prime-mover.

13. Induction motors of the slip-ring type are started by means of resistance connected to the rotor winding via slip-rings. The same resistors are also used to control the motor's speed on load.

14. Cage rotor type induction motors may be started

   (i) directly on to the line,
   (ii) by star-delta starter,
   (iii) by auto-transformer starter.

Their speed may be controlled by varying

   (i) supply frequency,
   (ii) by pole-changing,
   (iii) by pole amplitude modulation.

EXERCISES

1. Define the following induction machines in terms of the windings used and currents flowing in them:

   (i) cage type and (ii) slip ring type.

2. Using the two equations $e = Bl\omega r$ (V) and $F = Bli$ (N), describe briefly the operation of an induction motor.

3. Show how a rotating magnetic field is produced by three coils spaced at 120° to each other and fed from a three-phase a.c. supply. Calculate the speeds of the field when the supply frequency is 50 Hz and 60 Hz. Give answers in rad/s and rev/min.

Answers: 314 rad/s;   3000 rev/min;   375·4 rad/s;   3600 rev/min.

4. A developed cross-section of the stator of a three-phase, two-pole machine is shown in Fig. 8.26.

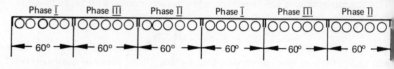

Fig. 8.26

Draw an mmf diagram at an instant when,

      (i) current in Phase I is maximum,
and   (ii) current in Phase II is zero.

Use it to show that such a winding fed from a three-phase supply produces a rotating mmf wave at synchronous speed.

If the frequency of the supply is 60 Hz calculate the speed of rotation of the mmf wave in rev/min.

How can the direction of rotation of the mmf wave be reversed?

Answer: 3600 rev/min.

5. (a) Three-phase windings mounted on a stator and carrying 50-Hz currents, produce a magnetic field rotating at 1500 rev/min.

Calculate the number of equivalent stator poles.

If a cage rotor placed inside the stator revolves first at (i) 1400 rev/min. and then (ii) 1580 rev/min describe the operating mode of the resulting induction machine for each rotor speed.

(b) Explain briefly why the induction machine cannot operate as a generator supplying an isolated resistive load.

Answers: (a) 4.

6. Define slip speed and slip.
Find the speed of rotation and frequency of the rotor currents in a three-phase 6-pole induction machine operating from a 60-Hz supply at the following slips:

      (i) +0·003;   (ii) +0·035;   (iii) −0·04;   (iv) +1

Indicate in what manner the machine is operating in each case.

Answers: (i) 1196·4, 0·18;   (ii) 1158, 2·1;   (iii) 1248, 2·4;   (iv) 0, 60.

7. Tabulate synchronous speeds for a series of induction machines having 2, 4, 6, 8, 10, 12 and 14 poles when they are connected to:

      (i) a 50-Hz supply,
      (ii) a 60-Hz supply,
      (iii) a 100-Hz supply.

Explain how the direction of rotation of the magnetic field can be reversed.

8. Deduce the expression for

      (i) the e.m.f. induced in each stator phase,
and   (ii) the e.m.f. induced in each rotor phase.

If the stator winding consists of fully pitched coils made up of 600 conductors per phase, with a spread factor of 0·92, whereas the rotor windings has 400 conductors per phase with a spread factor of 0·9, calculate the two e.m.f's at standstill when the supply frequency is 50 Hz and the flux per equivalent pole is 0·002 Wb.

What would be the rotor e.m.f. if the machine were operating at a slip of +0·04?

Answers: 122·5 V,   79·9 V,   32 V.

9. An 8-pole induction machine is connected to three-phase, 50-Hz, constant voltage busbars. Its standstill rotor voltage per phase is 110 V, its standstill rotor reactance per phase is 10 Ω and its rotor resistance per phase 0·8 Ω.

   (i) When the machine runs as a motor at 720 rev/min calculate slip, rotor frequency, rotor current and the torque imparted to the rotor by the magnetic field.

   (ii) When the machine is driven as an induction generator at 780 rev/min calculate the slip, frequency, rotor current and the torque of the prime-mover. Neglect all losses.

Answers: (i) 0·04,   2 Hz,   4·9 A,   1440 Nm.
          (ii) −0·04,   2 Hz,   4·9 A,   1440 Nm.

10. (a) Explain how a rotating magnetic flux is produced by the stator of a three-phase two pole induction motor, illustrating your answer in terms of

   (i) waveforms of stator current;
   (ii) simple diagrams of the stator winding assuming one slot/pole/phase.

  (b) (i) Derive a general expression for the speed of rotation of such a flux, and,
     (ii) Describe and illustrate how a motoring torque would be produced in a squirrel cage rotor.

11. A six-pole, three-phase, 440-V, 50-Hz induction motor develops 8 b.h.p. (5·97 kW) at 955 rev/min when the power factor is 0·85. Stator losses amount to 400 W and the frictional losses to 0·5 h.p. (373 W)

Calculate: (a) the slip,
          (b) rotor resistive loss,
          (c) input power to motor,
          (d) efficiency,
          (e) line current,

for the load condition described.

Answers: 0·045,   268·6 W,   7009·6 W,   0·85,   10·8 A.

12. A three-phase a.c. machine has three coil windings connected to six unmarked terminals. Describe the test which needs to be conducted to establish connection of windings to terminals.

If the machine produces 240 V at 50 Hz in each winding, describe how you would connect them correctly (a) in star; (b) in delta.

13. (a) Describe the principle of operation of a squirrel cage induction motor, giving reasons for its name.

(b) A three-phase, six-pole, 50 h.p. (37·3 kW), 400-V, 50-Hz squirrel cage induction motor runs at 935 rev/min when fully loaded. Its efficiency and power factor at this load are 88% and 0·8 respectively.

Calculate: (i) Synchronous speed of the rotating magnetic field,
          (ii) Percentage slip at full load,
          (iii) Line current taken by the motor,
          (iv) The output torque of the shaft

Answers: 1000 rev/min, 6·5%,   76·6 A,   381 Nm.

14. A motor-generator comprising a three-phase 50 Hz cage-type induction motor and a shunt-connected d.c. generator is to be installed. From the following data, supplied by the makers, determine the current which will be taken by the motor on full load.

Generator:	Terminal voltage	240 V
	Current	25 A
	Efficiency at full load	70%
Motor:	Efficiency	75%
	Power factor	0·8
	Line voltage	400 V

(C. & G.L.I. 57 Pt.II 1960)

Answer: 11·4 A.

15. State the essential components of the starting gear necessary for use with a slip-ring induction motor. Give a circuit diagram to show these components connected to the motor. Indicate the position of any safety devices. How would any of these components differ if used for speed control?

What factors govern the value and rating of starting resistance which must be inserted in the rotor circuit of a slip-ring induction motor?

(C. & G.L.I. 57 Pt.II 1967)

16. Describe, giving reasons in each case, the effect of increasing the rotor resistance of a three-phase induction motor on

(a) the starting torque,
(b) the starting current,
(c) the full load efficiency,
(d) the shape of the speed-torque characteristic.

(C. & G.L.I. 57 Pt.II 1968)

17. Describe TWO methods by which the speed of a three-phase induction motor may be controlled; give circuit diagrams. State the advantages and disadvantages of each method and quote an appropriate application.

(C. & G.L.I. 57 Pt.II 1968)

18. A three-phase cage-rotor induction motor is started 'direct-on-line' by a four push-button station.

When button A is pressed the motor starts and continues to run in the forward direction until the button B is pressed. When button C is pressed the motor starts and continues to run in the backward direction until the button B is pressed. If, however, the button D is pressed and held down in addition to the button A or C the motor will run only while the buttons are held down.

Draw a schematic connection diagram of a suitable contactor control system. Include the necessary interlocks.

(C. & G.L.I. 57 Pt.II 1969)

Three-phase synchronous machines

This chapter describes the theory and operation of three-phase synchronous machines. In common with all other types of electromagnetic machines, their ability to convert power in either direction is shown by means of the equation $T\omega = e_r i$. As generators, synchronous machines are used to produce practically all the electrical power in every country of the world, and as such they form the central core of every power station. When operated as convertors of electrical to mechanical energy, they are referred to as synchronous motors. Their application in industry however, is not as universal as that of induction motors because of their greater complexity and consequent greater cost.

9.1   Definition of synchronous machines

A synchronous machine is defined as having the following two main windings:

      (i) a distributed phase winding, carrying three-phase alternating currents,

and   (ii) a coil winding which is either distributed or concentrated and carries direct current.

In small machines, below approximately 50 kW rating, the coil winding is usually mounted on the salient poles of the stator, and the phase winding on the cylindrical rotor resulting in 'stator fed' types. In large and very large machines in the Megawatt range up to 800 MW and over, the stator carries a phase winding, and the rotor a coil winding, giving the 'rotor fed' type.

9.2   Construction of typical three-phase synchronous machines

Most large synchronous machines in use today, are of the 'rotor fed' type, and may be further subdivided in two main groups:

      (i) those with a *cylindrical* rotor, wound with a distributed coil winding over a part of its circumference,

and   (ii) those with a *salient pole* rotor, which have concentrated coils wound around each pole.

Sectioned views of a turbo-generator (Fig. 9.1) and hydro-generator (Fig. 9.2) show the construction of the first and second type respectively.

    The power rating of all machines is proportional to their volume. This is clearly seen from the two illustrations. The round rotor machine, which invariably has two

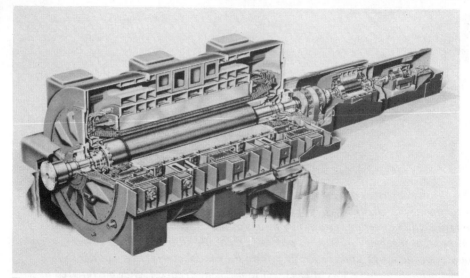

Fig. 9.1   Cutaway perspective of 300 MW turbo-generator, showing the
stator core, water cooled stator winding, direct hydrogen cooled
rotor, built-in hydrogen coolers and direct-driven exciter-set.

(*English Electric–A.E.I. Turbine Generators Ltd.*)

Fig. 9.2   A sectional model of a 66 667 kVA, 11 000 V, 3 phase, 50 Hz,
150 rev/min vertical hydro-generator. (*G.E.C. Machines Ltd.*)

poles and works at high speed, is small in diameter and therefore has large axial length. For example, its overall diameter may be two to three metres, whereas its length ten to twelve metres. In contrast a slow speed salient pole machine of similar rating which may have as many as 40 poles on its rotor, would be ten to fifteen metres in diameter and one to two metres along its shaft.

9.3   Simple explanation of operation of synchronous machine

Figure 9.3 shows the cross-section of a simplified synchronous machine.

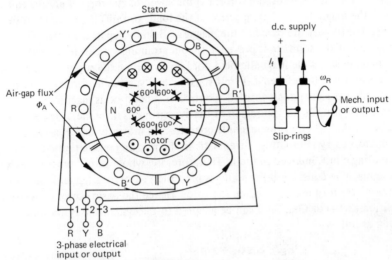

Fig. 9.3   Cross-section of simplified synchronous machine

The coil winding on the rotor is continuously fed from a d.c. source via brushes and slip-rings, or alternatively from rectifiers mounted on the rotating shaft. The direct current flowing in the conductors distributed around the rotor circumference produces an m.m.f. wave (see Chapter 6), which in turn establishes a magnetic flux in the air-gap, the distribution of which is nearly sinusoidal. The paths of 'average' magnetic flux lines are shown in Fig. 9.3.

When a mechanical prime-mover drives the rotor, the latter carries with it the magnetic flux, and in so doing, cuts the conductors mounted on the stator.

The three-phase windings are spaced at 120 degrees to each other and therefore are not cut simultaneously by the rotor flux, but at intervals equal to the time it takes the rotor to travel through one-third of its complete revolution. Thus the alternating e.m.f. induced in one phase is identical with that in the second and third but $120°$ out of phase with them. When the rotor's direction of rotation is anticlockwise as shown, then the e.m.f. induced in RR' leads that in YY' which in turn is in front of the BB' phase e.m.f. This sequence is regarded as positive.

If the rotation is reversed the sequence is RR', BB', and YY' and is considered negative.

The machine operates as a generator of three-phase alternating p.d. and when the load is connected to terminals 1, 2 and 3, three-phase alternating currents will flow.

The induced e.m.f.s are sinusoidal in waveform and their frequency depends on the speed of rotation of the prime-mover.

When the same machine is connected to a three-phase supply, each of the windings receives alternating current. These produce a rotating magnetic field in the air-gap which is equivalent to two magnetic poles, as explained in Chapter 6. Since the rotor has two magnetic poles produced by direct current, the pairs of opposite polarity 'lock in' together, and the rotor revolves in unison with the stator magnetic field. The machine then works as a motor, and delivers mechanical power at the shaft. Three points must be made: (i) the number of equivalent poles on the stator must be equal to the number of poles on the rotor; (ii) the speed of the rotor is always equal to that of the magnetic field irrespective of the load; and (iii) the rotor must be accelerated to near the speed of the magnetic field by separate means. This is because the weight of the solid rotor prevents it from attaining synchronous speed instantaneously as the magnetic field does on switching the supply to the stator.

This motor is therefore NOT 'self starting' and various methods of starting it are discussed in the paragraph entitled 'Control of synchronous machines'.

In considering an air-gap region of the synchronous machine, it is clear from the discussion that whether the machine operates as a motor or as a generator, its stator conductors carry alternating currents, lie in the rotating magnetic field, and have an alternating e.m.f. induced in them. Therefore, the whole stator experiences a torque. In the machine considered the stator is prevented from rotating and therefore an equal and opposite torque is exerted on the rotor. Hence the basic power conversion equation (3.14) of Chapter 3 can be adapted to the synchronous machine's air-gap region as follows:

$$3E_S I_S \cos \phi_S = T\omega_R \qquad\qquad (9.1)$$

where E_S = effective value of the e.m.f. induced in each phase winding,

$\quad\ \ I_S$ = effective value of the current flowing in each phase,

$\quad\ \ \phi_S$ = phase angle between phasors E_S and I_S,

$\quad\ \ T$ = average torque at which electro-mechanical conversion takes place

and $\quad \omega_R$ = constant speed of the rotor.

Factor 3 is included because the machine has three stator windings. The average quantities which replace instantaneous values in equation (3.14) are considered in detail in the succeeding paragraphs. It is further shown how expression (9.1) is modified when the whole machine is discussed.

9.4   Frequency and speed of synchronous machine

Equations deduced in Chapter 8 give the relationship between the frequency of the supply the number of equivalent poles and the speed of rotation of the magnetic field. Since the rotor of the synchronous motor must rotate at exactly the same speed as the magnetic field, and if:

$\qquad\qquad\qquad \omega_R$ = rotor speed in rad/s

and $\qquad\qquad\ \ n_R$ = rotor speed in rev/s

then the speed of synchronous motor is given by:

$$\omega_R = \omega_S = \frac{2\pi f}{p} \text{ (rad/s)} \tag{9.2}$$

or

$$n_R = n_S = \frac{f}{p} \text{ (rev/s)} \tag{9.3}$$

When the machine is operating as a generator, each stator conductor has one complete cycle of e.m.f. induced in it, when one pair of rotor poles passes under it. Thus in one revolution, the rotor with p pairs of poles induces p cycles. If the rotor rotates at ω_R rad/s then it makes $\omega_R/2\pi$ revolutions in one second, and the frequency of the e.m.f. induced in each stator conductor is

$$f = \frac{p\omega_R}{2\pi} \text{ (Hz)} \tag{9.4}$$

or in terms of n_R rev/sec

$$f = pn_R \text{ (Hz)} \tag{9.5}$$

Equations (9.2) and (9.3) are exactly the same as (9.4) and (9.5) respectively. Hence the relationship between *the frequency* of e.m.f.s and currents in phase windings and the *speed of the rotor is the same* whether a synchronous machine operates as a motor or as a generator (alternator).

9.5   Basis of phasor diagram and circuit representation

In Fig. 9.4(a) the cross-section of a 'stator fed' synchronous machine is shown at the instant when the induced e.m.f. in the RR$'$ coil is a maximum. The directions of e.m.f. in all three windings are shown by dots and crosses, for the anticlockwise rotation of the rotor. The magnetic flux due to m.m.f. F_A produced by an exciting current I_f flowing in a stator coil winding of N_f turns acts from left to right. Assuming sinusoidal flux in the air-gap the e.m.f.s in all three-phases will also be sinusoidal. All the quantities can therefore be represented by the phasors as shown in Fig. 9.4(b), with the m.m.f. and flux phasors lagging e.m.f. E_R by 90° as in the case of a pure reactor. In Fig. 9.4(c) the whole machine is drawn in a simplified form, where the three-phase windings as well as a d.c. coil winding are all represented by graphical symbols. The coils are shown connected in star, but may also be joined in delta. Synchronous machines, especially when operating as generators, are normally connected in STAR configuration, whereas motors are more often in delta. In Fig. 9.5 the cross-section, the phasor diagram and the circuit are drawn for a 'rotor fed' synchronous machine at the same instant of time as for a 'stator fed' machine. The rotor is again assumed to revolve in an anticlockwise direction, since that is the positive direction of rotation of the phasors. The consequences of the interchange of the windings between the rotor and the stator however, can be seen in Fig. 9.5(a) and are as follows:

(i) The m.m.f. F_A and the air-gap flux Φ_A due to it act from right to left and therefore must be shown *leading* the e.m.f. E_R by 90°,

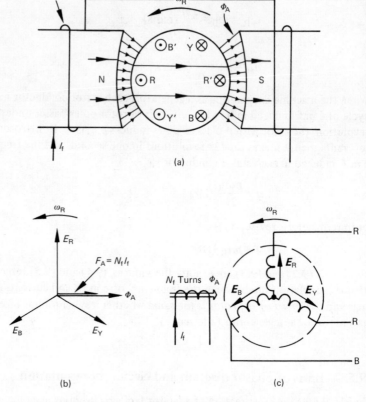

Fig. 9.4 'Stator fed' synchronous machine

and   (ii) the position of YY' and BB' windings must be interchanged to preserve the positive phase sequence of the induced e.m.f.'s (i.e. E_R leading E_Y, and E_Y leading E_B).

Since all the phase windings are identical, the analysis of the machine in terms of p.d.s. e.m.f.s and currents is usually done for one phase only, the other two having exactly the same values but 120° out of step with the considered phase.

From the point of view of magnetic fluxes however, the whole machine is considered because:

   (i) only *one* coil winding carries direct current and produces one exciting m.m.f.
and   (ii) the three-phase windings with alternating currents flowing in them, produce *ONE* rotating m.m.f. as seen in Chapter 6.

In this book only the cylindrical 'rotor fed' machines, that is those with uniform air-gap are considered. This type is used to develop the theory which nevertheless applies equally well to a 'stator fed' type.

The analysis of salient pole machines is slightly more complex due to non-uniform air-gap and is not given here.

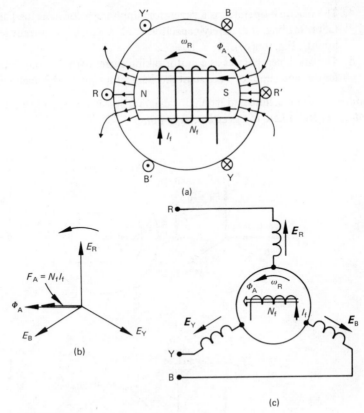

Fig. 9.5 'Rotor fed' synchronous machine

9.6 Simplified phasor diagrams for five conditions of operation of synchronous machine

In order to visualise the relationship between

    (i) phasor diagram
    (ii) magnetic flux inside a machine due to both the rotor and the stator m.m.f.s,
and (iii) the circuit representation of the machine for different conditions of operation,

the following five cases are illustrated:

1. The machine operates as a generator on no load (Fig. 9.6(a)).
2. The machine operates as a generator supplying a constant load at unity power factor (i.e. a purely resistive load in which voltage and current are in phase, Fig. 9.6(b)).
3. The machine operates as a generator supplying a constant load at zero power factor lagging, (i.e. purely inductive load in which the current lags the voltage by $90°$, Fig. 9.6(c)).

4. The machine operates as a generator supplying a constant load at zero power factor leading, (i.e. purely capacitive load in which the current leads the voltage by 90°, Fig. 9.6(d)).

5. The machine operates as a motor taking current from a source at unity power factor, and supplies a constant mechanical torque to the load, (Fig. 9.6(e)).

For case 1 the machine and its phasor diagram are shown at an instant when the *voltage in RR′ phase is a positive maximum.*

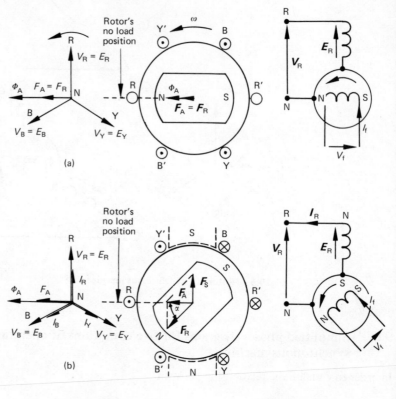

(a)

(b)

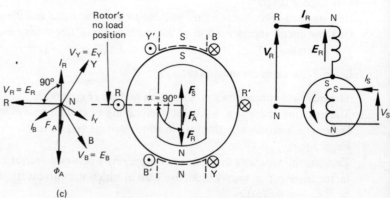

(c)

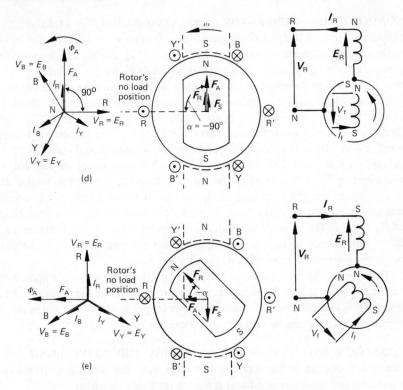

Fig. 9.6   (a) Generator on no-load (b) Generator supplying a load at unity p.f. (c) Generator supplying a load at zero p.f. lagging (d) Generator supplying a load at zero p.f. leading (e) Motor working at unity power factor. Five conditions of operation of a synchronous machine

For cases 2 to 5 the instant chosen is that at which the *current in RR' phase is a positive maximum* and therefore currents in the other two phases are half maximum and negative (the instant is shown by the line XX' in Fig. 6·16(a). Chapter 6). Furthermore the following simplifications are also made:

   (i) the machine has two poles only,

and   (ii) the stator windings are assumed to be purely inductive coils with negligible resistances and leakage reactances.

CONDITION (i) – GENERATOR WITHOUT A LOAD

The phasor diagram in Fig. 9.6(a) is drawn for an instant at which the generated e.m.f. of rotation, E_R in the RR' phase winding is a positive maximum. Since the winding is on open circuit and behaves as an inductive coil, its terminal voltage V_R is equal to E_R and the air-gap flux Φ_A due to rotor m.m.f. F_A leads it by 90°.

At that instant it is clear that the position of the rotor's magnetic poles must be central with respect to RR' conductors as explained in Chapter 6, and by a cross-sectional sketch of the machine.

The circuit diagram of the machine in Fig. 9.6(a) shows the relative positions of the rotor coil winding and the RR' phase winding.

CONDITION (ii) – GENERATOR SUPPLYING RESISTIVE LOAD

Starting again with the phasor diagram in Fig. 9.6(b); $V_R = E_R$ since the winding has negligible resistance and leakage reactance. V_R and I_R are in phase because the load power factor is unity, and the air-gap flux Φ_A leads E_R by $90°$ as before. However, Φ_A is now due to the resultant of the two m.m.f.s, the first F_R produced by the rotor and the second F_S produced by the stator load currents flowing in all three phases. The resultant m.m.f. F_A generates Φ_A which rotates in an anticlockwise direction and at the instant shown cuts the RR′ conductors inducing in them an e.m.f. E_R in such a direction that the current flows into R′ and out of R conductor (Fleming's R.H. rule). The currents in YY′ and BB′ conductors are then half maximum and negative. Their directions are shown by dots and crosses in Fig. 9.6(b). Thus the m.m.f. due to the phase winding has an upward direction which may be considered due to two equivalent stator poles shown by dotted lines. F_A is the geometric sum of F_S and F_R. The rotor poles producing F_R must therefore be strengthened and their position must be moved forward in an anticlockwise direction by an angle α in order to preserve the same value of flux Φ_A.

Considering the machine as made up of two rotor and two stator poles it may be said that when the machine works as a generator, the prime-mover drives the rotor poles which exert a 'push' on the magnetic poles produced by the stator currents – thereby converting mechanical into electrical energy.

CONDITION (iii) – GENERATOR SUPPLYING INDUCTIVE LOAD

The phasor diagram in Fig. 9.6(c) is drawn for an instant when the current I_R is a maximum and therefore its phasor is shown vertically upwards.

An inductive load demands that E_R, equal to V_R, is $90°$ ahead of I_R. Therefore F_A generating Φ_A, which leads E_R by $90°$ is in direct opposition to the stator m.m.f. F_S as seen in the cross-sectional sketch of the machine (Fig. 9.6(c)).

The rotor m.m.f. F_R must now be increased by the amount equal to F_S if the original value of Φ_A is to be maintained. The angle α becomes $90°$ and the effect of stator m.m.f. F_S referred to as armature reaction is purely demagnetising. Furthermore, no torque is developed between the stator and rotor magnetic poles as they are in line with each other, and therefore the prime-mover need not supply any power other than that required to overcome the rotational losses. The generator does not deliver electrical power to an inductive load, although it does supply it with the voltage and the current, i.e., the reactive volt-amperes.

The circuit diagram shows, as before, the relative positions of rotor and the RR′ winding of the machine.

CONDITION (iv) – GENERATOR SUPPLYING CAPACITIVE LOAD

In Fig. 9.6(d) the phasor diagram, the sketch of the machine's cross-section, and the circuit diagram, show that when a capacitive load is connected to a generator, then:

(a) the angle α is $-90°$, the negative sign indicating clockwise shift of the rotor from its no-load position;

(b) the stator and rotor m.m.f.s are in line and arithmetically add to F_A, which generates an air-gap flux Φ_A. Thus F_R must be *reduced* to maintain the original value of F_A;

(c) the armature reaction is thus purely magnetising;

(d) no torque is developed between the rotor and stator poles and no electrical power is delivered to the load;

(e) the generator delivers reactive volt-amperes only.

CONDITION (v) – SYNCHRONOUS MOTOR SUPPLYING MECHANICAL LOAD

To illustrate this mode of operation, the motor is assumed to draw a current from the supply at unity power factor, although both leading and lagging values are possible, as described in subsequent paragraphs. The diagrams in Fig. 9.6(e) show that the direction of the currents in the stator phase windings are reversed and therefore so are the polarities of the stator equivalent poles, as compared with Fig. 9.6(b).

Furthermore, the load angle α becomes negative with respect to case (ii) and the stator poles exert the push on the rotor. The latter experiences the torque and electrical power is converted to mechanical power.

In all five cases described so far, the position of the rotor with respect to its own 'no-load' position, considered at the *same* instant of time during its rotation, was found to be different. This position may be characterised by the value, of an angle α, termed the *load angle*, which may be either positive or negative.

When the machine works as a separate generator, that is not connected to any network, then,

(a) at no load, $\alpha = 0$,

(b) at load power factors lagging between unity and zero, α lies between 0 and $+90°$,

(c) at load power factors leading between unity and zero, α lies between 0 and $-90°$.

When the machine works as a synchronous motor it must always be connected to the supply network. This fact has much greater consequences than the reversal of stator current only and necessitates closer investigation, which is given later in this chapter.

Here, however, the comparison of case (ii) and (v) shows clearly that if current and voltage are equal for generator and motor operation, then α is equal but positive in the former and negative in the latter case.

9.7   E.M.F. of rotation and Open Circuit characteristic of a synchronous machine

Whether a machine of either cylindrical or salient pole construction operates as a generator or as a synchronous motor, the three-phase windings are 'cut' by the sinusoidally distributed magnetic flux Φ_A rotating relative to them in the air-gap. Since all three windings are identical it is only necessary to consider one, the e.m.f.s being the same in the other two but out of phase with each other by $120°$. Furthermore, the expression for the e.m.f. is the same for both conditions of operation, and is deduced in Chapter 6. Hence,

$$E_S = 4.44 \; k_s k_p \Phi_A N f \text{(V)} \tag{9.6}$$

where E_S = r.m.s. value of the e.m.f. in each stator phase

Φ_A = the air-gap flux per pole due to rotor and stator m.m.f.s in Wb

N = the number of turns in each phase connected in series

f = the frequency in Hz, which is related to the speed of rotation by an expression (9.4) and (9.5)

k_s = the winding spread or distribution coefficient

k_p = the pitch or span coefficient for each turn.

When the machine is unloaded, i.e. no current flows in the phase windings, the air-gap flux Φ_A is due to the direct current flowing in the coil winding alone. Just as in a d.c. machine, this excitation current is designated by I_f.

It is usual to relate the e.m.f. E_S to the exciting current I_f or more commonly to the exciting ampere-turns (current I_f multiplied by the turns N_f of the coil winding) graphically as shown in Fig. 9.7.

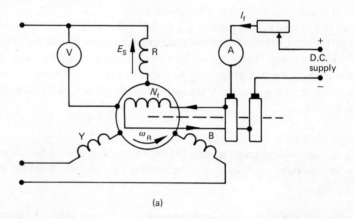

(a)

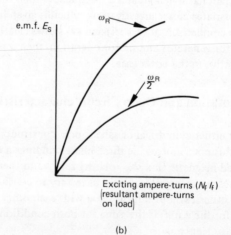

(b)

Fig. 9.7   Circuit diagram and Open Circuit characteristic for a synchronous machine

The curves known as 'Open Circuit characteristics' are obtained experimentally by running a machine as a generator at a series of constant speeds. The machine is not loaded, the field current I_f is varied from zero to a maximum and the generated e.m.f. per phase E_S is measured. The results (E_S) are plotted against $(I_f N_f)$ where N_f is the number of turns of the exciting winding mounted on the rotor. The circuit diagram, with the necessary instruments, is shown in Fig. 9.7.

When either a mechanical or electrical load is applied to the machine, the e.m.f. E_S is due to the air-gap flux Φ_A which is produced by two m.m.f.s; that due to the rotor F_R, and F_S due to the three load currents flowing in the stator turns of all three-phase windings.

If the resultant ampere-turns F_A are found, then the e.m.f. of rotation E_S can be read off directly from the Open Circuit characteristic, just as if the machine were unloaded. The resultant is thus treated as if it were due to the rotor alone.

It must be stressed that the load current in a synchronous machine has a very much greater effect on the excitation than in a d.c. machine. This is so because in addition to its magnitude, its power factor also plays a major part in determining the final value of the air-gap flux Φ_A. It is therefore never possible to neglect the effect of armature reaction except under almost no load conditions.

9.8   Torque of synchronous machine

Whether the synchronous machine is loaded as a motor or as a generator, the phase windings carry alternating currents, and lie within the rotating magnetic field. The frequencies of the currents and the speed of the magnetic field are always related by equation $\omega_R = 2\pi f/p$ and therefore each conductor experiences a force according to expression $F = Bli$. In a 'rotor fed' machine, the whole of the stator experiences a torque, but since the stator is prevented from rotating, an equal and opposite torque is exerted on the rotor. At constant speed and loading, this torque must be overcome by the mechanical prime-mover if the machine is to generate electrical power.

In the motor case, the torque produces rotation and hence mechanical power.

In either case the expression for the torque is the same and is deduced from equation (9.1), i.e.,

$$T = \frac{3}{\omega_R} E_S I_S \cos \phi_S \text{ (Nm)} \qquad (9.7)$$

Substituting expression (9.2) for ω_R the equation becomes

$$T = \frac{3p}{2\pi f} E_S I_S \cos \phi_S \text{ (Nm)} \qquad (9.8)$$

If all losses in the machine are neglected, then T is the prime-mover's torque driving the machine as a generator, or the mechanical torque developed by the synchronous motor demanded by the load.

QUESTION 1. A three-phase synchronous machine, connected to 50 Hz bus-bars, has the following particulars:

Number of rotor poles	6
Number of stator turns per phase	100
Magnetic flux per pole under full load conditions	0·00115 Wb
Winding spread factor	0·96
Winding pitch factor	0·985
Full load current in each phase	50 A
Full load power factor	0·8 lagging

Calculate for full load conditions,

(i) the rotor's speed in rad/s and rev/s
(ii) the e.m.f. of rotation generated in each phase winding
(iii) the power being converted by the machine,
(iv) the torque at which this conversion takes place.

ANSWER.

(i) Using expression (9.2).

$$\omega_R = \frac{2\pi f}{p} = \frac{2\pi 50}{3} = \underline{104\cdot 7} \text{ rad/s}$$

and (9.3)

$$n_R = \frac{f}{p} = \frac{50}{3} = \underline{16\cdot 67} \text{ rev/s}$$

(ii)   E.M.F. per phase $= 4\cdot 44 \, k_s k_p \cdot \Phi_A N f$
$$= 4\cdot 44 \times 0\cdot 96 \times 0\cdot 985 \times 0\cdot 00115 \times 100 \times 50$$
$$= \underline{240 \text{ V}} \quad \text{♭} \text{ } ᴥ ⊾ \cdot |ᴙ ✓$$

(iii)   Power converted $= 3E_S I_S \cos\phi_S$
$$= 3 \times 240 \times 50 \times 0\cdot 8 = \underline{28\,800 \text{ W}}$$

(iv) $$T = \frac{28\,800}{104\cdot 7} = \underline{276 \text{ Nm}}$$

9.9   Equivalent circuit of a synchronous machine

The rotor fed synchronous machine has already been represented by the graphical symbols in Fig. 9.5(c). This representation is now extended further so that the important characteristics of each winding are shown by separate elements. The elements for each phase winding are:

(i) a pure source of an e.m.f. of rotation E_S (r.m.s. value) which is wholly governed by the magnitude and speed of rotation of the air-gap flux Φ_A.
(ii) a pure ohmic resistance R
(iii) a pure leakage reactance X, which is due to leakage flux linking the conductors of the phase winding only.

If L is the leakage inductance then $X = 2\pi f L$. The rotor coil winding, which carries direct current needs only to be represented by its resistance R_f.

The modified circuit shown in Fig. 9.8(a) and (b) is known as the equivalent circuit of a synchronous machine. The circuit diagram in Fig. 9.8(a) is marked for generator operation, whereas that in Fig. 9.8(b) for motor operation. Furthermore only one phase of the machine is shown since the other two are identical with it in every respect except their physical location on the stator's circumference. It must also be remembered that the air-gap flux Φ_A is produced by the m.m.f. F_A which is the sum of stator m.m.f. F_S and the rotor m.m.f. F_R. F_S is due to current flowing through all three-phase windings whereas F_R is due to field current through the turns of one coil winding only.

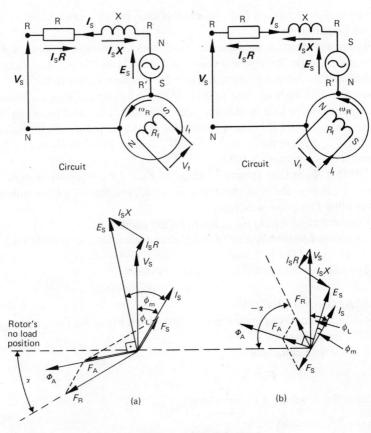

Fig. 9.8   (a) Generator (b) Motor. Equivalent circuit and
phasor diagram of a synchronous machine

9.10   Circuit equations and phasor diagrams for a synchronous machine

GENERATOR
Applying Kirchhoff's voltage law to the circuit in Fig. 9.8(a) the following two equations are obtained:

for the field coil circuit

$$V_f = I_f R_f \tag{9.9}$$

and for the RR$'$ phase of the stator winding (moving clockwise round the circuit):

$$V_S + I_S R + I_S X - E_S = 0 \qquad (9.10)$$

The third equation relating the m.m.f.s follows from the fact that the air-gap flux Φ_A is caused by the stator m.m.f. F_S and the rotor m.m.f. F_R. Hence if F_A is the resultant m.m.f. then:

$$F_A = F_S + F_R \qquad (9.11)$$

The equations (9.10) and (9.11) contain phasor quantities and therefore must be solved graphically.

The phasor diagram for a machine working as a generator is drawn in Fig. 9.8(a) at an instant when the terminal voltage V_S in RR$'$ phase is a positive maximum. Its construction is as follows. Having drawn V_S vertically upwards, the current I_S is drawn at a load angle ϕ_L. The magnitude and the phase angle of the current are determined by the load on the generator. The load is assumed to be lagging. The p.d. across the resistance R being in phase with I_S and the p.d. across the leakage reactance X, leading the current I_S by 90° are then added to V_S to give E_S as demanded by equation (9.10).

The flux Φ_A is then drawn 90° ahead of E_S and F_A is shown in phase with it. The m.m.f. F_S is now drawn in phase with I_S, which produces it jointly with the current in the other two phase windings.

From equation (9.11) F_R is found by the parallelogram method of phasor addition. The combined phasor diagram in Fig. 9.8(a) enables the solutions of both equations (9.10) and (9.11) to be found.

The diagram is sometimes termed the 'Potier diagram'.

SYNCHRONOUS MOTOR

Equations for motor operation are obtained from Fig. 9.8(b) in the same manner as that for the generator.

For the field coil circuit

$$V_f = I_f R_f \qquad (9.9)$$

which is the same as for the generator.

For the RR$'$ phase of the stator winding

$$V_S - I_S R - I_S X - E_S = 0 \qquad (9.12)$$

and for the m.m.f.'s

$$F_A = F_R + F_S \qquad (9.11)$$

which again is the same as in the generator case.

The solution of the last two equations by means of the composite phasor diagram is shown in Fig. 9.8(b).

Assuming lagging power factor of the load, the p.d.s $I_S R$ and $I_S X$ are now subtracted from V_S to give E_S. The air-gap flux Φ_A together with F_A is shown 90° ahead of E_S.

The m.m.f. F_S however must now be drawn in opposition to the current I_S because the latter's direction is reversed, i.e. it is now drawn from the supply. F_R is obtained as before by application of the equation (9.11).

The load angle α becomes negative, showing that the 'stator's magnetic poles' exert the push on the rotor, thereby converting electrical to mechanical energy.

The phasor diagrams in Fig. 9.8(a) and (b) are fuller versions of those shown in Fig. 9.6.

QUESTION 2. A 3-phase 11-kV, 10-MVA, 50-Hz, 2-pole, rotor-fed, star-connected synchronous machine, run as a generator on open circuit, gave the following results:

E.M.F. of rotation per phase in kV } 0 1·72 3·5 4·77 5·73 6·36 6·88 7·3 7·62

Rotor exciting m.m.f. in A } 0 50   100 150 200 250 300 350 400

Winding resistance per phase   = 0·1 Ω
Winding leakage reactance per phase = 1 Ω
Stator ampere-turns on full load   = 200 A

Find (i) e.m.f. per stator phase,
      (ii) resultant m.m.f. producing an air-gap flux,
      (iii) rotor m.m.f.,
      (iv) load angle,

when the machine is (a) supplying 10 MVA to the load at the line voltage of 11 kV and 0·8 power factor lagging, and (b) drawing 10 MVA from the 11 kV bus-bars at 0·8 power factor leading.

ANSWER.   Phase voltage $V_S = \dfrac{11}{\sqrt{3}} = 6·36$ kV

          Phase current $I_S = \dfrac{10 \times 10^6}{3 \times 6.36 \times 10^3} = 524$ A

          Voltage drop across $R = 524 \times 0·1 = 52·4$ V.
          Voltage drop across $X = 524 \times 1 = 524$ V.

Phasor diagrams are drawn to scale as explained in paragraph 9.10.

Generator

          From phasor diagram $E_S = 6700$ V
          From O.C. curve m.m.f. at 6700 V = 280 A
          From phasor diagram rotor m.m.f. = 436 A
          Load angle $\alpha = 23·2°$

Motor

          From phasor diagram $E_S = 5900$ V
          From O.C. curve m.m.f. at 5900 V = 220 A
          From phasor diagram rotor m.m.f. = 372 A
          Load angle $\alpha = -22·9°$

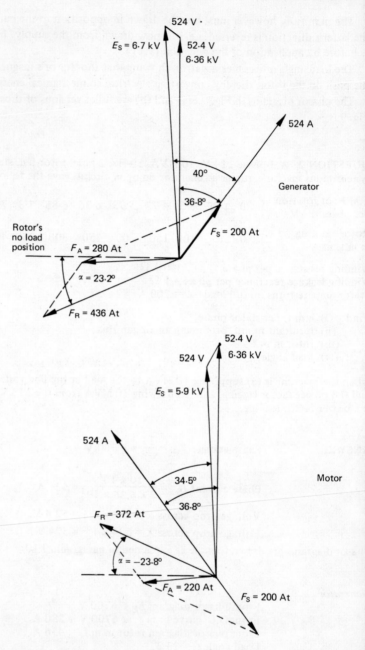

Scales: 1 kV = 1 cm, 100 A = 1 cm, 100 At = 1 cm

Fig. 9.9 Phasor diagrams for Question 2

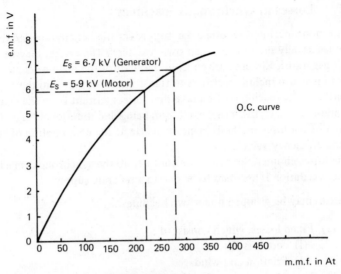

Fig. 9.10   O.C. characteristic for Question 2

9.11   Power equations for synchronous machines

When the circuit and phasor diagrams of the previous paragraphs are considered from the point of view of power conversion, then the following equations are obtained:

For a generator replace by:

$$3V_S I_S \cos \phi_L = 3E_S I_S \cos \phi_S - 3I_S^2 R$$

and for a motor replace by:

$$3V_S I_S \cos \phi_L = 3E_S I_S \cos \phi_S + 3I_S^2 R$$

but $T\omega_R = 3E_S I_S \cos \phi_S$ by equation (9.1)

therefore substituting and combining the two equations,

$$3V_S I_S \cos \phi_L = T\omega_R \pm 3I_S^2 R \qquad\qquad (9.13)$$

is obtained.

The factor 3 is included because the machine has three stator windings, and in the diagram the voltage, current, resistance and power factor values are those of one phase.

It is obvious from the above that the power input, whether electrical *OR* mechanical exceeds the output by the stator resistive loss amounting to $3I_S^2 R$ watts. The rotational losses are neglected at this point, and so also is the resistive loss in the rotor field winding. The latter is assumed to be supplied from the separate source of electrical energy.

9.12   Losses in synchronous machines

Synchronous machines require a d.c. supply for their excitation and therefore larger machines usually incorporate their own 'exciters'. The exciter may be either a d.c. or an a.c. generator. Either is driven from the shaft of the main machine. An a.c. exciter must also include rectifiers to transform its output into the required direct current. Smaller machines often take their direct current from the main winding via solid state rectifiers, the initial excitation being due to residual magnetism.

Each of the three methods mentioned affects the computation of losses and therefore the efficiency values.

The losses enumerated below are confined to the synchronous machine itself and the d.c. excitation is assumed to be due to a separate supply.

The losses may be grouped under two headings:

    (a) Fixed losses, which consist of
        (i) core loss P_c,
       (ii) Friction and windage.
and (b) Variable losses composed from:
        (i) Resistive loss $3I_S^2 R$, in the three-phase windings,
      (ii) Resistive loss $I_f^2 R_f$, in the coil winding,
     (iii) Stray loss.

The synchronous machine operates at absolutely constant speed and the e.m.f. of rotation is reasonably constant under normal operating conditions. Thus the air-gap flux is approximately constant and so hysteresis and eddy current losses remain unaltered. The core loss is confined to either stator or rotor whichever one carries the three-phase winding. There is no core loss in the component carrying the coil winding since it *always* rotates at the *same* speed as the magnetic flux. The variable losses depend on the load, which determines the value of the load current. The stray loss is difficult to determine and is neglected in this book. It is due to extra leakage fluxes when the load and field currents are increased. This loss therefore increases the 'fixed' core loss at higher loads. The fixed losses are obtained from the no load test, in which the machine is driven as a generator with an excitation adjusted to give the no load e.m.f. it would require at rated full load. The value of this e.m.f. is calculated beforehand from the Open Circuit characteristic and the phasor diagram. The output of the prime-mover is thus wholly dissipated in overcoming the generator core, friction, and windage loss.

Variable losses may be obtained by calculation using measured values of R and R_f. As in all machines the resistances should be measured 'hot' and include terminals, and/or slip rings and brushes.

9.13   Efficiency of a synchronous machine

The per unit efficiency can be calculated from the equation given in Chapter 1.

The power factor may either be leading or lagging for generator or motor operation.

QUESTION 3. Calculate the full load efficiency of a synchronous generator from the following information:

Rating   30 MVA at power factor 0·8 lagging
Line Voltage: 11 kV
No. of phases: 3
Stator winding: Star connected with 0·2 Ω per phase.
Rotor field winding loss at full load: 100 kW.

The core, friction and windage loss obtained from an Open Circuit test with excitation to give calculated e.m.f. at normal rated load is 600 kW.

ANSWER.

$$\text{Stator full load current per phase} = \frac{30 \times 10^6}{\sqrt{3} \times 11 \times 10^3} = 1575 \text{ A.}$$

Full load output $= 30 \times 10^6 \times 0·8 = 24 \times 10^6 \text{ W}$

Losses: Core, Friction & Windage $= 0·6 \times 10^6 \text{ W}$

Rotor resistive loss $= 0·1 \times 10^6 \text{ W}$

Stator resistive loss $= 3 \times (1575)^2 \times 0·2 = 0·148$
                                          $= 0·148 \times 10^6 \text{ W}$

Total $= 0·848 \times 10^6 \text{ W}$

Therefore input to the generator $= 24·848 \times 10^6 \text{ W}$

$$\text{Efficiency} = \frac{24 \times 10^6}{24·848 \times 10^6} = 0·966 \text{ p.u.}$$

9.14   Simplified equivalent circuit – synchronous reactance

In paragraphs 9.9 and 9.10 it was shown that the air-gap flux Φ_A is due to rotor and stator m.m.f.s. Furthermore, the stator m.m.f. varies not only in magnitude but also in phase, when the value and power factor of the stator current changes. This effect referred to as armature reaction, has such a large influence on the machine's operation that it cannot be neglected except at very light loads. Although the armature reaction cannot be omitted, nevertheless it is possible to simplify the machine and still allow for the effect of the stator's m.m.f.

The procedure for this is as follows:

(i) The stator m.m.f., which is shown in Fig. 9.8 (a) and (b) is omitted;
(ii) It is assumed that the air-gap flux, Φ_A is now due to the rotor's m.m.f. F_R only;
(iii) The flux Φ_A generates, by virtue of the rotor's rotation, a new and fictitious e.m.f. in the stator winding. This e.m.f. is designated by E_O and lags Φ_A by 90°.
(iv) The difference between E_O and the terminal voltage V_S is due to (a) potential difference across stator resistance R and equals $I_S R$ as before, and (b) potential difference across the new and fictitious stator's winding reactance X_s. This reactance is termed synchronous reactance and is much larger than the actual leakage reactance X which it includes.

The modified phasor diagram is shown in Fig. 9.11(a) and should be compared with Fig. 9.8(a). Figure 9.11(b) shows the simplified equivalent circuit of the machine working as a generator. The circuit corresponds to the new phasor diagram.

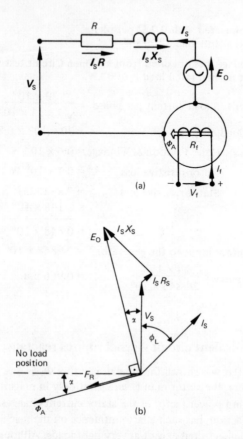

Fig. 9.11   Simplified equivalent circuit and phasor diagram of a synchronous generator

As an exercise the reader should draw similar diagrams for synchronous motor operation.

The simplified equivalent circuit introduces *two* fictitious values:

(1) A new e.m.f. of rotation E_O which is larger than E_S,

and   (2) A new reactance X_s, which is also larger than X.

These two quantities in effect preserve the actual value and position of the rotor m.m.f. with respect to the actual positions of the phasors for terminal voltage V_S and current I_S. They allow for the armature reaction (i.e. stator m.m.f.) by increasing the reactance of the stator windings.

The value of E_O is related to the rotor m.m.f. by an Open-Circuit characteristic and either can be read off the curve when the other is known.

The justification of such a step lies in,

(i) a much simpler representation of the machine,

(ii) an easier view of the machine's behaviour which becomes apparent when its operation from a large supply network is considered,

and (iii) reasonable agreement between calculated and actual results – provided the value of X_s is chosen with care.

9.15   Estimation of synchronous reactance from test results

The value of synchronous reactance may be obtained from the following test. A synchronous machine is run as a generator at the rated speed and with its terminals short-circuited through three ammeters. The circuit is shown in Fig. 9.12(a).

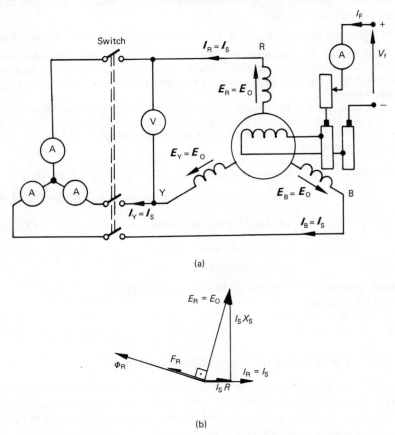

(a)

(b)

Fig. 9.12 Circuit and phasor diagram for short-circuit test on a synchronous machine

The rotor current I_f is slowly increased until the rotor m.m.f. F_R produces E_O sufficient to drive the current I_S at which the synchronous reactance is required. The value of I_S is taken as an average reading of the three ammeters. It is clear from the phasor diagram in Fig. 9.12(b) that the whole of the generated e.m.f. E_O is used up in overcoming the potential difference across R and X_s.

The terminal voltage V_S is zero because the potential differences across the ammeters are negligibly small. Without altering the speed and the value of the field current I_f the switch is opened. The voltmeter now reads the e.m.f. E_O which must be the same as when the generator was on short-circuit.
Hence:

$$(R^2 + X_s^2)^{\frac{1}{2}} = \frac{E_O}{I_S}$$

and

$$X_s = \left(\left(\frac{E_O}{I_S} \right) - R^2 \right)^{\frac{1}{2}} \qquad (9.14)$$

The value of the resistance R for one phase winding is obtained by any of the usual methods of measurement.

The combined expression $\sqrt{R^2 + X_s^2} = Z_s$ is known as 'the Synchronous Impedance'.

QUESTION 4. An excitation of a three-phase, two-pole, star-connected synchronous generator with its terminals short-circuited was adjusted until the full load current of 1000 A/phase was obtained. The same excitation on open circuit produces 4400 V/phase.

Calculate the synchronous reactance of each winding if its resistance is $0 \cdot 6$ Ω/phase.

Use this value to estimate E_O and the load angle α when the machine, working as a generator, supplies the full load current at the line voltage of 11 kV and a lagging power factor of $0 \cdot 8$.

ANSWER.   Synchronous impedance $Z_s = \dfrac{E_O}{I_S} = \dfrac{4400}{1000} = 4 \cdot 4$ Ω

synchronous reactance $X_s = (4 \cdot 4^2 - 0 \cdot 6^2)^{\frac{1}{2}} = \sqrt{19} = 4 \cdot 36$ Ω

Terminal voltage per phase $= \dfrac{11\,000}{\sqrt{3}}$ $= \underline{6360 \text{ V}}$

$I_S R = 1000 \times 0 \cdot 6 = 600$ V

$I_S X_s = 1000 \times 4 \cdot 36 = 4360$ V

Phasor diagram is constructed in the usual way in Fig. 9.13.
From the diagram:
E_O/phase $= 10 \cdot 5$ k V
and $\alpha = 18 \cdot 3°$

9.16   The meaning of 'Large supply system'

The tremendous and continually growing demand for electrical power in every country has led to interconnections between a very large number of power stations, through transmission and distribution lines. All the generators therefore work in parallel and form one gigantic synchronous machine.

In the same way all the users of electrical energy connected to this supply constitute a very large, variable load.

Each additional or incoming machine, whether working as a generator or as a motor is therefore connected in parallel with the whole system. The above is shown pictorially in Fig. 9.14.

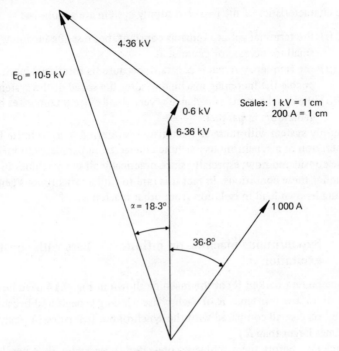

Fig. 9.13 Phasor diagram for Question 4

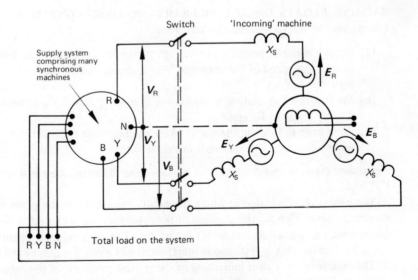

Fig. 9.14 Large power system with 'incoming' machine

The characteristics of the depicted supply system are as follows:

(i) the terminal voltage remains constant, because the incoming machine is too small to increase or decrease it,

(ii) the frequency remains constant, because its rotational inertia is too large to enable the incoming machine to alter the speed of the system,

and (iii) the synchronous impedance is very small since it comprises hundreds of generators in parallel.

A supply system with these characteristics is referred to as 'Infinite Bus-bars'. The operation of a synchronous machine connected in parallel with infinite bus-bars must be considered now; especially since practically all the machines in use today work under these conditions. In fact it is rare to find a synchronous generator supplying its own load in isolation from other machines.

9.17   Synchronous machine on infinite bus-bars with constant excitation

Only one phase (marked R) of the machine shown in Fig. 9.14 need be considered. Furthermore, the resistance R of each phase winding is neglected because its value is usually very small compared with the synchronous reactance (X_s may be 100 or more times larger than R).

In practice, before the machine is connected to the network, a number of conditions have to be observed. These conditions and the process of connecting the machine to the system is described in a paragraph on 'control'. The procedure followed is termed synchronisation.

Assuming that the machine has been synchronised and the switch closed, three different modes of operation occur:

MACHINE FLOATS ON THE BUS-BARS – NO-LOAD CONDITION
The machine is said to float on the bars when,

(a) the prime-mover provides just enough power for the rotor to be driven at exactly the speed of the magnetic field as demanded by the relation $n_R = f/p$ (rev/s.)

(b) the machine's excitation is adjusted so that rotor m.m.f. F_R produces E_O exactly equal to V_S and

(c) no current I_S flows in the stator phase winding, that is E_O is in phase with V_S and no resultant voltage exists in the circuit.

The phasor diagram drawn 'inside' the cross-section of the machine is shown in Fig. 9.15.

The instant chosen is that at which the supply voltage V_S in the phase RR' is a maximum. Since $V_S = E_O$ the position of the rotor flux Φ_A is central to the RR' conductors, i.e. the middle conductor is 'cut' by a flux of maximum density. The e.m.f. E_O lags the flux by 90° and is therefore drawn along the axis of the RR' phase.

The diagram gives a clear indication of the relative positions of the rotor with respect to the stator, during the former's rotation at this particular instant of time.

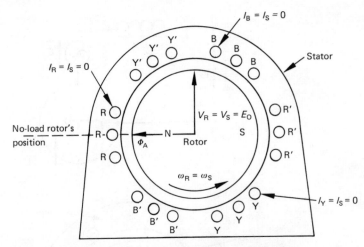

Fig. 9.15   Phasor diagram for machine floating on the bus-bars

Since the prime-mover maintains E_O in phase with V_S no current and therefore no power is either drawn or delivered to the supply network. All the losses occurring in the machine are thus supplied by the prime-mover.

MACHINE SUPPLIES POWER TO THE BUS-BARS – GENERATOR OPERATION

A single generator delivers electrical power to its own load as soon as the latter is connected to it, *provided* the mechanical prime-mover maintains its original speed. The prime-mover must thus deliver greater power to the generator according to the demands of the latter's load. This is normally arranged by providing the mechanical prime-mover with a speed sensitive governor. The governor's function is to deliver a greater amount of fuel to it as soon as the speed of the system begins to drop.

When the machine is connected to the infinite supply network together with its large load, neither load switching nor speed changes are observable.

The initiative must come from the prime-mover's side. More fuel is admitted to the prime-mover which in turn generates larger torque and pushes the rotor forward with respect to its no load position. This results in an electrical power being fed into the network. When the increased driving power is exactly balanced by the electrical power delivered, the rotor's forward movement stops and its speed continues to be the same as before, but in a new position.

In Fig. 9.16(a) the circuit diagram marked with voltages and currents and the phasor diagram in Fig. 9.16(b) are shown. The phasor diagram is drawn for the same instant as in previous case, namely when the instantaneous value of the supply voltage V_S in the Red phase is a maximum.

Let the angle by which the rotor is advanced be α, then Φ_A is forward with respect to its no-load position, and the generated e.m.f. E_O is also α degrees in advance of V_S. E_O no longer balances the V_S and the difference between them appears across the synchronous reactance i.e.,

$$V_S + I_S X_s = E_O$$

or

$$I_S X_s = E_O - V_S \tag{9.15}$$

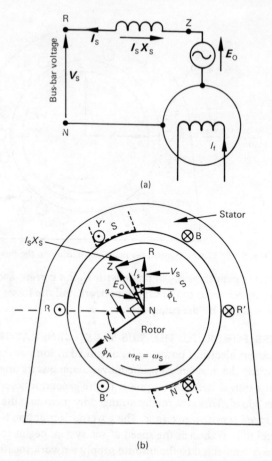

Fig. 9.16   (a) Circuit (b) Phasor diagram. Generator on infinite bus-bars

A graphical solution of (9.15) is obtained by means of the phasor diagram. The value of I_S and its position relative to the potential difference across X_s is deduced from the knowledge that I_S must lag $I_S X_s$ by $90°$, because the synchronous reactance has a purely inductive effect.

In this case the current I_S leads V_S by a small angle, and the power from one phase delivered to the network is $V_S I_S \cos \phi_L$, where $\cos \phi_L$ is a leading power factor.

The value of angle α and consequently the amount of electrical power delivered to the bus-bars is controlled by the amount of mechanical power developed by the prime-mover.

MACHINE DRAWS POWER FROM THE BUS-BARS – MOTOR OPERATION

When the driving engine is disconnected from the synchronous machine floating on the bus-bars, its rotor continues to revolve at exactly the same speed, i.e. $\omega_R = \omega_S$.

If mechanical load is now applied to its shaft, the machine develops a driving torque and the mechanical power output is obtained from it. This occurs because the rotor falls back by an angle $(-\alpha)$ and the phasor E_O lags V_S by the same angle.

The circuit and the phasor diagrams are shown in Fig. 9.17.

Voltage equation is obtained by applying Kirchhoff's voltage law:

$$V_S - I_S X_s - E_O = 0$$

or

$$I_S X_s = V_S - E_O \tag{9.16}$$

The potential difference across X_s is now the phasor difference between V_S and E_O because the current I_S is drawn from the bus-bars. The power per phase supplied to the machine is given by $V_S I_S \cos\phi_L$ where $\cos\phi_L$ is a lagging power factor. The position of I_S is at right angles to the voltage $I_S X_s$ as before. When the mechanical load is increased, the rotor falls further back increasing the negative angle $(-\alpha)$. This increases the magnitude of $I_S X_s$ and therefore the current I_S. The speed of rotation however remains the same, only the position of the rotor varies with respect to that at no-load.

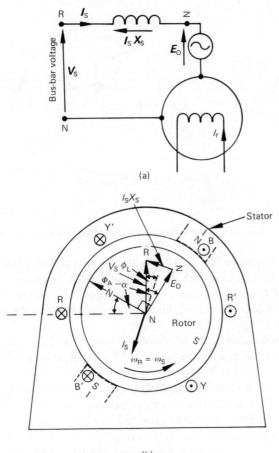

(a)

(b)

Fig. 9.17 (a) Circuit (b) Phasor diagram. Synchronous motor on infinite bus-bars

9.18   Synchronous machine on infinite bus-bars with constant power

Again three cases are examined. These are,

(i) the machine operates as a generator supplying power to the bus-bars;
(ii) the machine works as a motor driving a mechanical load;
(iii) the machine works as an unloaded motor.

CASE (i)

A generator, the circuit of which is shown in Fig. 9.16(a) delivers electrical power to the supply network equal to $3V_S I_S \cos \phi_L$, when the e.m.f. of rotation E_O is set at a predetermined value. Here the effect of varying the excitation which changes E_O is considered, assuming that the power output is kept constant. To fulfill this condition the prime-mover's driving torque is so adjusted that it does not allow an increase or decrease in electrical input to the bus-bars.

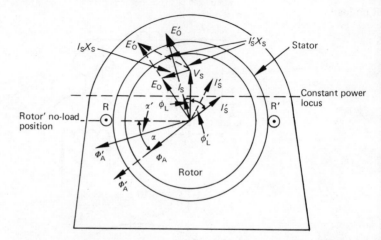

Fig. 9.18   **Phasor diagram for generator with increased excitation**

In the phasor diagram of Fig. 9.18 the locus of constant power is shown by the horizontal dotted line, because with the supply voltage V_S remaining constant, $I_S \cos \phi_L$ (the projection of I_S on to V_S) must also remain constant. Hence as long as the tip of the I_S phasor does not exceed this line the power input to the bars will remain unaltered.

Let the excitation current I_f be increased to a new value. This increases Φ_A, which in turn causes E_O to rise to a new value marked E_O' and shown by a dotted line. The rotor position is unaltered at first because of its inertia. The new e.m.f. E_O' increases the p.d. across X_s which in turn increases I_S' (dotted line). However I_S' now lies above the horizontal line. This means that greater power would now be supplied to the bars. The prime-mover however is not set to provide larger driving torque, and the extra power demand produces a retarding torque on the rotor. The rotor falls back, decreasing angle α, until the tip of I_S' falls to the dotted line. The new position of equilibrium is shown by continuous phasors for Φ_A', E_O' and I_S'. By comparing the positions and the magnitudes with the original Φ_A, E_O and I_S, it is observed that the power factor of the current I_S with respect to V_S has now changed from lead to lag, and the angle α has been slightly reduced.

Thus the excitation (I_f current through N_f turns) controls the power factor and slightly alters the load angle α, since it cannot affect the terminal voltage V_S.

In the case of the generator, the increase of I_f makes the power factor more lagging and reduces the load angle α.

CASE (ii)

The reader should draw a phasor diagram for the machine working as a motor when its excitation is increased, starting with the diagram in Fig. 9.17(a). The conclusions from this exercise are the same as for a generator, except that an excitation increase changes the power factor from *lag* to *lead*, i.e. opposite to that of the previous case. The load angle $(-\alpha)$ is also slightly reduced just as $+\alpha$ was reduced for generator operation.

CASE (iii)

When the synchronous machine is allowed to rotate without any mechanical load attached to its shaft and when the excitation current I_f is adjusted to a large value, then E_O is in excess of V_S and the phasor diagram is as shown in Fig. 9.19.

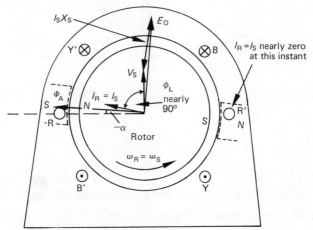

Fig. 9.19   Phasor diagram for synchronous condenser

The power taken from the bus-bars is very small for it is only needed to supply the machine's losses, hence angle $(-\alpha)$ is very small and E_O is practically in line with V_S. The p.d. across the synchronous reactance is thus nearly the arithmetic difference between E_O and V_S and the stator current I_S is almost 90° ahead of the bus-bar voltage V_S.

The machine behaves like a capacitor. Furthermore, the current I_S can be controlled at will by altering the excitation, making the machine, in fact, a variable capacitor.

The machine so operated is used to improve the power factor of various installations and is called a *'synchronous compensator'*.

It should also be noticed that my making E_O smaller than V_S the current I_S will lag V_S by almost $90°$ and thus the machine becomes an inductor with high 'inductive reactance to resistance ratio', and may be used in that capacity also. However, in industrial applications the former method is most commonly used.

9.19   Synchronous machine on infinite bus-bars – Power and torque expressions

It is shown in paragraph 9.3 that the total power converted in a synchronous machine is given by $3E_SI_S \cos \phi_S$ (W). This expression is modified for a synchronous machine in parallel with infinite bus-bars by reference to the phasor diagram in Fig. 9.16(b) and Fig. 9.17(b).

E_S is replaced by E_O, which is now the e.m.f. generated in the stator winding and ϕ_S by $(\alpha - \phi_L)$, since the resistance R is neglected.

Thus the power generated equals the power received by the bus-bars for a generator and vice-versa for the motor. In equation form:

$$3E_OI_S \cos (\alpha - \phi_L) = 3V_SI_S \cos \phi_L \qquad (9.17)$$

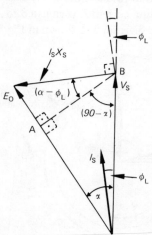

Fig. 9.20   **Synchronous generator phasor diagram**

Re-drawing the phasor diagram of Fig. 9.16 in Fig. 9.20, the equation (9.17) can be expressed in a different form as follows:

The length of dotted line AB $= V_S \sin \alpha = I_SX_s \cos (\alpha - \phi_L)$

$$\therefore I_S \cos (\alpha - \phi_L) = \frac{V_S \sin \alpha}{X_s}$$

Substituting for $I_S \cos (\alpha - \phi_L)$ in equation (9.17) gives:

$$\text{Power} = 3\frac{E_OV_S}{X_s} \sin \alpha \text{ (W)} \qquad (9.18)$$

and the torque T is:

$$T = \frac{3}{\omega_R} \cdot \frac{E_OV_S}{X_s} \sin \alpha \text{ (Nm)} \qquad (9.19)$$

The equations apply to a synchronous machine working either as

    (i) a generator,

or (ii) a motor.

and both the power and the torque obtainable are the values at which an electro-mechanical power conversion takes place. The actual values may be calculated by making due allowance for rotational losses. The stator resistive loss however, is not taken into account even so, because equations (9.18) and (9.19) are deduced with resistance R neglected.

QUESTION 5. A three-phase two-pole, 11-kV, 100-MVA, 50-Hz, rotor fed, star connected machine has a synchronous reactance of 0·5 Ω per phase. If its excitation is set to give an e.m.f. of rotation E_O 10% higher than the bus-bar voltage to which it is connected, find

    (i) the power factor,
    (ii) the load angle,
    (iii) the power,
    (iv) the torque,

when converting full load MVA as (a) a generator
                                  (b) a synchronous motor.

ANSWER.   Phase voltage $V_S = \dfrac{11\,000}{\sqrt{3}} = 6360$ V

$$E_O = 6360 + 0{\cdot}1 \times 6360 = 6996 \text{ V}$$

Full load current $I_S = \dfrac{100 \times 10^6}{3 \times 6360} = 5250$ A

P.D. across synchronous reactance $5250 \times 0{\cdot}5 = 2625$ V

Using these values the phasor diagrams are drawn to scale in Fig. 9.21.

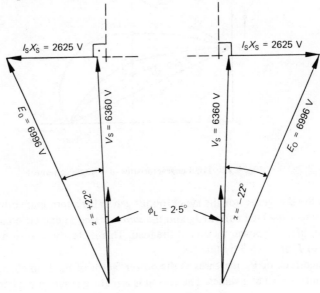

Scale 1 cm = 1 000 V

Generator   Motor

Fig. 9.21

From the diagrams:

$$\begin{array}{ll}
\textit{generator} & \textit{motor} \\
\phi_L = 2\tfrac{1}{2}° \text{ lag} & \phi_L = 2\tfrac{1}{2}° \text{ lead} \\
\text{p.f.} = \cos 2\tfrac{1}{2}° = 0\cdot999 \text{ lagging} & \text{p.f.} = 0\cdot999 \text{ leading} \\
\alpha = +22° & \alpha = -22°
\end{array}$$

$$\text{Power} = 3 \times \frac{6996 \times 6360}{0\cdot5} \sin 22° = \underline{99\cdot9 \text{ MW}}$$

or

$$3 \times 6360 \times 5250 \times \cos 2\tfrac{1}{2}° = \underline{99\cdot9 \text{ MW}}$$

$$\text{Torque} = \frac{99\cdot9 \times 10^6}{3000 \times 2\pi/60} = \underline{318\,000 \text{ Nm}}$$

9.20 Characteristics of synchronous machines

SINGLE GENERATOR

The system consists of the mechanical engine, driving a synchronous machine as a generator, which in turn has an electrical load connected to its terminals. The prime-mover is provided with a speed-sensitive governor so that automatically any increase of the load on the generator causes it to increase the mechanical driving torque and so maintain a constant speed of the arrangement. The load characteristics are thus the curves relating phase terminal voltage V_S to the magnitude of the load current I_S supplied by the stator windings. The curves are shown in Fig. 9.22.

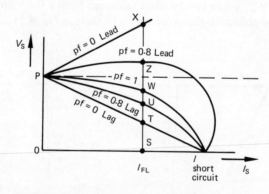

Fig. 9.22 Load characteristics of a single generator

Unlike the d.c. generator the synchronous generator's terminal voltage has many values for the *same* load current I_S and the *same* setting of excitation ampere-turns $(I_f N_f)$, but differing power factors of the load. This is clearly seen from the Fig. 9.22 at a value of I_S indicated by line SX.

The magnitude of V_S increases as the power factor of the load changes from lagging through unity to leading values. The reason is armature reaction explained in detail in paragraph 9.6.

The ratio of voltage drop between no load and full load, to its full load value is referred to as *regulation*, and is a measure of the generator's capacity to maintain the

terminal voltage at different demands put upon it by the load, e.g. the regulation for
the machine operating at point U would be given by:

$$\text{Regulation} = \frac{OP - SU}{SU} \text{ p.u.}$$

where all the ordinates are measured in volts.

SYNCHRONOUS MOTOR

All synchronous motors operate from a supply of constant frequency and voltage.
Since they are able to develop torque only at synchronous speed, which is determined
by the frequency, hence their speed/torque characteristic is a horizontal straight line
shown in Fig. 9.23.

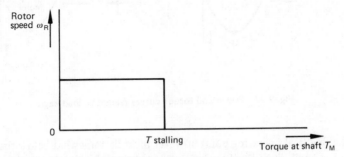

Fig. 9.23   Speed/torque characteristic of a synchronous motor

The speed is absolutely constant over the full operating range, until the load torque
exceeds the value which the machine is able to supply. At this point the motor stalls
and stops. This torque is referred to as the *STALLING TORQUE*.

SYNCHRONOUS MACHINE ON INFINITE BUS-BARS

The voltage and frequency of a large supply network remain constant and are not
changed by the behaviour of the synchronous machine connected to it. Furthermore
the machine, whether operating as a generator or as a motor rotates at absolutely
constant speed. In these circumstances the operating characteristics of the synchronous
generator cannot be expressed by a graph of terminal voltage plotted against the load
current. Similarly the torque plotted against the speed for a motor gives a straight line
as shown already and in that form does not provide the information normally needed.

In paragraph 9.19 the expressions relating power and torque to the rotor load
angle α were deduced, and when depicted in a graphical form they serve as the
operating characteristics for a synchronous machine working in parallel with 'Infinite
bus-bars'. Figure 9.24(a) shows the power and Fig. 9.24(c) the torque plotted against
the angle α.

In previous work the angle was assumed to be positive for a generator and negative
for a machine working as a motor. In consequence the power delivered to the
supply network is now regarded as positive, whereas that drawn from it, as negative.
Similarly the torque due to the mechanical prime mover is positive and that produced
by the synchronous motor negative.

Looking at both characteristics it is clear that the maximum positive or negative power and torque occur at $\alpha = \pm 90°$. This means that greater driving torque on the generator shaft or greater mechanical load on the motor shaft than this value will result in smaller output or input. Hence this is the static limit of stability, i.e. the point beyond which the generator's speed will increase above synchronous value, and the motor's will fall off to zero.

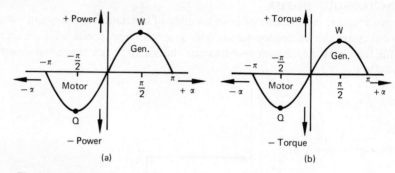

Fig. 9.24   Power and torque curves related to load angle

The practical operating point however must lie somewhat below that maximum to allow for small, short lived load variation, without causing the machine to break its magnetic link and 'fall out of synchronism'.

The two characteristics discussed are drawn only for *one* constant setting of excitation which determines the value of E_O, and in this case the limits of the operating torque are marked by points Q and W.

9.21   Control of synchronous machines

The control of a machine is described under the following headings:

(i) Control of excitation.
(ii) Control of power input or output.
(iii) Synchronising of generators with the supply.
(iv) Damping of the rotor oscillations, and
(v) Starting of synchronous motors.

9.22   Control of excitation

Almost all synchronous machines are built with their own means of producing direct current for excitation.

The conventional arrangement is shown in a schematic form in Fig. 9.25.

The machine carries on its own shaft the main and pilot exciters. Both are d.c. generators; the first feeding the field of a synchronous machine and the second the field of the first. The output voltage of the main d.c. generator is varied by adjusting

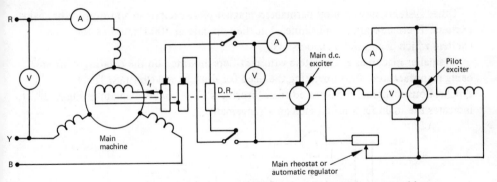

Fig. 9.25 Conventional excitation system for synchronous machine

its excitation by means of the rheostat, which can either be hand or automatically operated.

In smaller machines the pilot exciter is omitted and the main exciter field is shunt connected. This arrangement, however, is not sensitive or quick acting when changes of the field current are required by the main machine.

At the present time the development of solid state rectifiers has brought about the introduction of a.c. exciters. Figure 9.26 shows the circuit of a large synchronous machine using these in place of the conventional d.c. generators.

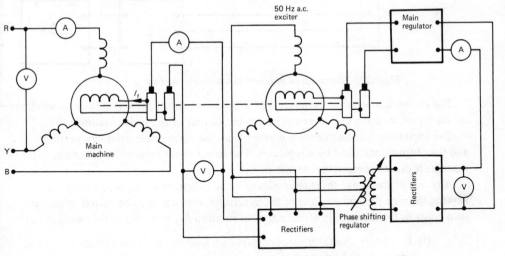

Fig. 9.26 Excitation system using a.c. exciters and silicon rectifiers

The output of an a.c. exciter is rectified and fed through the slip-rings to the main rotor coil winding. The exciter field is taken from its own a.c. output via a phase shifting induction regulator and separate set of rectifiers. The direct current so obtained is controlled by the main regulator which may employ an amplidyne set. This excitation is suitable for very large synchronous machines rated in hundreds of MVA.

Other systems may employ permanent magnet pilot exciters in addition to the main exciters. These also are a.c. machines, but they operate at 400 Hz, unlike the main exciters which generate 50 Hz outputs.

In smaller machines a.c. exciters with rectifiers mounted on the rotating shaft are being manufactured, thus obviating the need for brushes and sliprings. These synchronous machines are referred to as 'brushless'. The circuit diagram in Fig. 9.27 indicates the main features of such an arrangement.

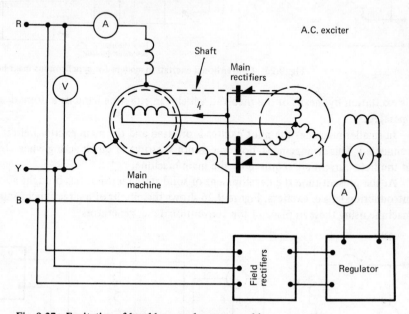

Fig. 9.27   Excitation of brushless synchronous machine

The exciter's three-phase winding is now mounted on the rotor and the coil winding on the stator, which is the reverse arrangement to that of the main machine.

The excitation for the exciter is derived from the main phase windings, rectified and the output controlled by a regulator. The initial direct current for the main machine is due to residual magnetism of the a.c. exciter.

Any one of the above three examples of excitation enables the field current I_f flowing through the coil winding of the synchronous machine to be varied, which in turn controls the magnitude of the e.m.f. of rotation E_O and has the following effects:

(i) In a single generator supplying its own load the terminal voltage V_S can be adjusted to the desired value.

(ii) In a generator connected to infinite bus-bars the power factor of the current *delivered* to the network, can be adjusted.

and   (iii) in a motor the power factor of the current *taken from* the network can be controlled.

In addition the magnitude of E_O determines the maximum value of the power and consequently the torque to which the machine can be loaded either as a motor or as a generator.

This is seen from the equations (9.18) and (9.19) and the graphs shown in Fig. 9.28.

There is an infinite number of power/angle characteristics between the limits of minimum and maximum values to which I_f can be set.

Each curve reaches maximum at $\alpha = \pm 90°$. It is therefore clear that the excitation predetermines the point at which the machine falls out of synchronism. That point increases with increased excitation i.e. the higher the value of E_O the stronger the magnetic link between the rotor poles and the equivalent stator poles.

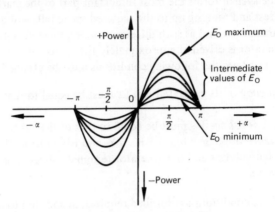

Fig. 9.28   Family of power/load angle curves

9.23   Control of power input or output

GENERATOR

The amount of electrical power delivered to a load or to infinite bus-bars by a generator is controlled by how much mechanical power the prime-mover is set to produce.

The difference between a generator operating singly and that connected to infinite bus-bars lies in the initiation of the process.

In a separate generator the 'switching in' of an electrical load causes slight drop of speed. The engine governor preset for normal speed detects the drop, admits more fuel to the prime-mover to restore the revolutions and thereby increases the driving power of the engine. Without the governor, the prime mover could not increase its power and therefore no amount of 'switching in' would increase electrical energy output.

In a generator connected to infinite bus-bars the initiative must come either from the operator or from an automatic device which is set to the pre-determined value of mechanical power to be converted. The fixed driving torque thus advances generator's rotor to a value of the load angle α consistent with the electrical power fed into the bars and equal to the mechanical power less losses.

2.6

MOTOR

In this case the amount of the mechanical power delivered is controlled by the magnitude of the load. The larger the load, the larger the load angle ($-\alpha$) by which the rotor falls back relative to its no load position, and the greater the amount of electrical power drawn from the supply. It is seen that in all cases, therefore, the control is confined to the 'mechanical' side of the system.

9.24   Synchronising a generator

Since most generators are part of a large electrical network the connection of each new unit to the system forms the most important part of the starting procedure. The starting from rest and running up to the required speed falls within the control of the mechanical prime-mover and as such differs according to what type of engine is used. Once the generator is driven at approximately the correct speed, and before it can be joined to the system, the following conditions must be attained.

  (i) the *frequency* of its generated e.m.f.s must be equal to that of the supply network,
  (ii) the *terminal voltage* V_S must be equal to that of the bus-bars,
  (iii) the *sequence* of phase e.m.f.'s RR', YY' and BB' must be the same for both,
and (iv) the bus-bar voltages and the generator terminal voltages must be *in step* i.e. *in phase.*

 To check that these conditions are fulfilled requires, in addition to ammeters and voltmeters, an instrument called a synchroscope. In its simplest form the synchroscope consists of three ordinary lamps. Figure 9.29 shows a circuit diagram with all the necessary instruments to achieve synchronisation. The procedure is as follows.

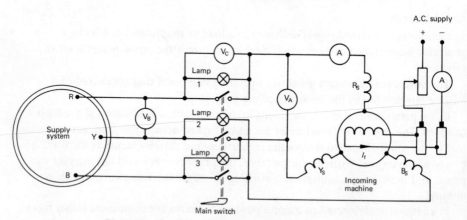

Fig. 9.29 Circuit to synchronise a generator to the supply

 The generator's speed is increased to near the value given by $n_R = f/p$ (rev/s.) The field current is increased until V_A reads the same voltage as V_B. The lamps connected across the poles of the main connecting switch should now slowly dim in turn. If they do not the sequence of phase e.m.f.s is incorrect and either the generator's rotation must be reversed or two line connections interchanged. At an instant when lamp 1 is out, the voltmeter V_C reads zero and all the e.m.f.s are in phase. This is the moment to close the switch. The lamps 2 and 3 should at that time be of equal brightness.

Figure 9.30 shows phasor diagrams of supply and generator voltages.

The generator's speed is assumed to be very slightly greater than that of the supply equivalent poles. The lamp 1 has the smallest voltage across it, whereas lamp 2 has the highest. As the generator phasors approach near to supply phasors, the voltage across lamp 1 decreases to zero, indicating phase coincidence. This condition does not last long and the switch must be closed quickly just before the middle of the dark period of lamp 1. The machine is now floating on the bas-bars and ready to take up the load either as a generator or, if its prime-mover is disconnected, as a motor.

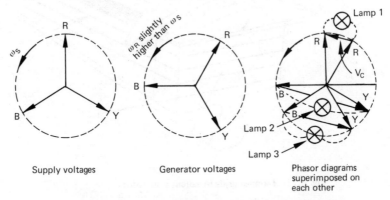

| Supply voltages | Generator voltages | Phasor diagrams superimposed on each other |

**Fig. 9.30   Phasor diagrams for instant just before complete synchronism
(a) Supply voltages (b) Generator voltages (c) Phasor diagrams
superimposed on each other**

In electric power stations more sophisticated equipment is used than that indicated in the Fig. 9.29. The three lamps are invariably backed up by a dial type synchroscope, which consists of three iron vanes spaced at $120°$ apart and mounted on the spindle. The spindle is placed inside the stator with three coils which are fed by the difference of bus-bar and generator voltages. The difference is proportional to (f supply $- f$ generator).

When very large machines are being synchronised the whole procedure is done automatically obviating the risk of error by an operator.

9.25   Damping of rotor oscillations

A synchronous machine connected in parallel with the network is joined to it by means of a magnetic link between the rotor and the stator poles. Under steady state conditions the rotor revolves at $\omega_R = 2\pi f/p$ (rad/s) and at a definite load angle ($\pm \alpha$). But if the load is changed or the rotors position otherwise disturbed requiring a change of α, then the rotor moves forward and backward about its new position because of its own inertia. These oscillations are superimposed upon its rotary movement and must decrease quickly. If instead they increase in magnitude then the machine will fall out of synchronism by breaking the magnetic link.

To damp down the oscillations, especially in slow speed, salient pole machines, a second winding is added to the rotor. The winding is cage type, mounted in slots cut in pole faces and shorted at both ends of the rotor. This winding is usually referred to as 'damper grids' and is shown in Fig. 9.31. Its action is as follows.

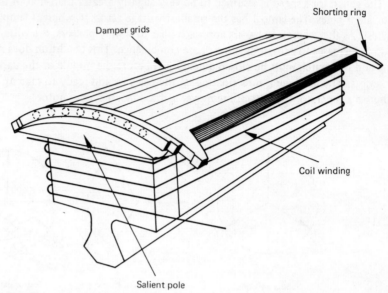

Fig. 9.31   **Damper grids in salient pole rotor**

When the rotor begins to oscillate about its position of equilibrium, each conductor of the damping winding is cut by the air-gap flux; first in one then in the opposite direction. Reversing e.m.f.s are thus induced and drive eddy currents through the grid conductors and their short-circuiting rings. The currents produce $I^2 R$ power loss in the whole winding thus absorbing the energy of the swing and reducing it to zero.

In cylindrical rotor machines, the rotor is usually solid and eddy currents are set up near its surface and produce the same damping effect as the grids in machines constructed with salient poles.

9.26   Starting of synchronous motors

It has already been stated that a synchronous motor develops a torque only at synchronous speed when the rotor poles due to direct current in the coil winding lock with stator equivalent poles due to alternating currents in three-phase windings.

It is therefore necessary to run up the machine to synchronous or near synchronous speed by other means. The methods used are:

    (i) pony induction motor

and (ii) starting as an induction motor.

The first method employs a small induction motor usually with one pair of poles less than the main machine to allow the latter to reach its synchronous speed. At this speed the d.c. supply is switched on to the rotor coil winding and synchronising torque is developed. The supply to the pony motor is then disconnected. This method is used for large motors and for synchronous compensators. An example of one such scheme is shown in Fig. 9.32.

At starting the switch Y and the field circuit breaker are both open and the main switch is closed. The supply is thus applied to the stator of the pony motor via the windings of the main machine. The windings act as series reactors and limit the starting current of the induction motor. When the speed attains synchronous value, the d.c. switch and the star point switch Y are both closed. The direct current magnetises the rotor, whereas switch Y connects the stator windings in star and at the same time short-circuits the pony motor. The main machine now operates as a synchronous motor, with the pony motor rotor revolving inside the de-energised stator.

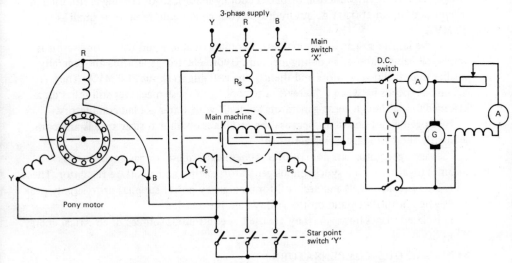

Fig. 9.32   Pony motor method of starting a synchronous motor

The second method makes use of damper grids in salient pole machines or the eddy currents induced in solid rotor of the cylindrical type motors. The supply is switched on to the three-phase windings, which produce a rotating magnetic field.

The coil winding is de-energised and the damper grids behave as a cage winding of an induction motor. When the speed of the rotor reaches its maximum sub-synchronous speed, direct current is fed into the coil winding and the rotor pulls into step. This occurs because of the slow pulsations of the synchronising torque produced as the rotor poles slip past the stator equivalent poles.

When it is necessary to limit the initial current, the machine is started by any of the methods used for cage type induction motors, i.e. the following may be employed:

    (i) series reactors
    (ii) star-delta switching
or (iii) auto-transformers.

9.27   Applications

SYNCHRONOUS GENERATORS

As the demand for electric power continues its growth, synchronous generators are being designed in ever larger sizes. Construction of the machines is largely defined by the prime-mover characteristics; the high speed of steam and gas turbines calls for cylindrical rotor design whereas the low speed of oil or gas engines and water turbines requires salient pole rotors.

Steam turbine generators are now manufactured mainly for power stations in sizes of up to 1000 MVA and operate at speeds of 25-60 rev/s. Units below 100 MVA are made for industrial plants. These increases in the ratings of individual units have been achieved by improvements in: magnetic materials, conductors, and insulating materials but principally by introduction of better cooling methods. Air cooling is still used in machines of up to 100 MVA, but hydrogen cooling is found in units as small as 30 MVA.

For the largest size an additional direct cooling of stator and rotor conductors is employed using water as a cooling medium. Hydro-electric generators are generally smaller than turbo-generators and their rating extends to about 300 MVA. They operate at speeds between 1-20 rev/s. The cooling of these machines still relies on air as a medium, although some generators are built with water cooled conductors.

Hydrogen is not necessary because the machines operate at slow speeds and consequently their windage losses are small.

Packaged-generator sets consist of a prime-mover (diesel or gas engine or gas turbine) together with a synchronous generator, exciter and a voltage regulator. These units are generally small and are used for emergency power supplies or to supply peak loads when normal demand on the grid system is exceeded.

In isolated places however, they are used as continuous sources of electrical energy.

SYNCHRONOUS COMPENSATORS

Synchronous compensators are machines operated continuously as unloaded synchronous motors connected to the grid system in order to enable transmission line circuits to transmit maximum active power under conditions of heavy loading at lagging power factors and to facilitate voltage control under conditions of light loadings at leading power factors.

Normally these machines operate either over-excited as capacitors (condensers) during heavy active load transfer, or under-excited as reactors during light loads. Machines of 40 MVA and over have been built for these duties.

SYNCHRONOUS MOTORS

Synchronous motors can be applied to many loads which can be driven satisfactorily by induction motors. However, the advantages of the synchronous motor are (1) higher efficiency, (2) leading power factor which can be adjusted to improve the over-all power factor of an installation, and (3) adaptability to large frame, low-speed applications because of large air-gaps.

Disadvantages are (1) necessity for an excitation source and control equipment, and (2) greater maintenance costs.

From the point of view of cost, the synchronous motors are cheaper than squirrel cage motors if the rating exceeds 45 kW per revolution per second (1 hp/rpm). This, of course, is a very rough approximation and the choice of the motor has to be evaluated by considering power factor, energy cost, hours of operation, and so on.

Thus at 50-60 rev/s synchronous motors may be used at 1500 kW and above and are the first choice above 4500 kW. In speed range of 8 rev/s to 20 rev/s and above 500 kW synchronous motors are generally employed.

The motors drive water pumps, wood-grinders, compressors rubber mills, steel mills, mine ventilating fans, etc.

SUMMARY

1. A synchronous machine is defined as having:

      (i) a distributed phase winding carrying alternating currents

and (ii) a coil winding carrying direct current.

Either of the two windings can be placed on the rotor or the stator. The rotor with the coil winding may be of 'salient pole' or cylindrical construction.

    A third, cage type, winding is sometimes placed in the slots cut in the faces of the salient pole rotor to reduce oscillations. It is termed a damper grid winding.

2. Power conversion equation at the machine's air-gap:

$$3E_S I_S \cos \phi_S = T\omega_R \tag{9.1}$$

where E_S = effective value of e.m.f. per phase

      I_S = effective value of current per phase

      ϕ_S = phase angle between E_S and I_S

      T = average torque at which electro-mechanical conversion takes place

      ω_R = the speed of the rotor.

3. Speed of rotation is given by:

$$\omega_R = \omega_S = \frac{2\pi f}{p} \,(\text{rad/s}) \tag{9.2}$$

or

$$n_R = n_S = \frac{f}{p} \,(\text{rev/s}) \tag{9.3}$$

4. Frequency is given by:

$$f = \frac{p\omega_R}{2\pi} \,(\text{Hz}) \tag{9.4}$$

or

$$f = pn_R \,(\text{Hz}) \tag{9.5}$$

5. E.m.f. of rotation per phase is given by:

$$E_S = 4 \cdot 44 \, k_s k_p \, \Phi_A Nf \,(\text{V}) \tag{9.6}$$

where E_S = r.m.s. value of the e.m.f. in each phase

      Φ_A = air-gap flux due to both windings

      N = number of turns in each phase connected in series

      f = frequency

      k_s = spread factor

      k_p = pitch factor.

6. Torque is given by:

$$T = \frac{3p}{2\pi f} E_S I_S \cos \phi_S \text{ (Nm)} \tag{9.8}$$

7. Circuit equations for:

(i) *generator*

field coil $\qquad V_f = I_f R_f$ $\hfill$ (9.9)

for phase winding per one phase

$$V_S + I_S R + I_S X - E_S = 0 \tag{9.10}$$

for m.m.f.'s $\quad F_A = F_S + F_R$ $\hfill$ (9.11)

(ii) *motor*

for field coil and m.m.f.s as for a generator.
for phase winding per phase

$$V_S - I_S R - I_S X - E_S = 0 \tag{9.12}$$

equations other than (9.9) must be solved by phasor methods.

8. The power equation is given by

$$3 V_S I_S \cos \phi_L = T \omega_R \pm 3 I_S^2 R \tag{9.13}$$

where a minus sign applies to a generator, and a plus sign applies to a motor.

9. Losses in synchronous machines are classified as:

    (a) Fixed losses

        (i) core loss P_c
        (ii) friction and windage loss C

and   (b) Variable losses

        (i) resistive loss $3 I_S^2 R$ in the three-phase winding
        (ii) resistive loss $I_f^2 R_f$ in the coil winding
     and (iii) stray loss.

10. Synchronous machine is usually represented by a simplified equivalent circuit which introduces 'synchronous reactance' concept.

Synchronous reactance X_s is a term which combines

    (i) leakage reactance

and   (ii) the effect of armature reaction, in one value.

$$X_s = \left(\left(\frac{E_O}{I_S} \right)^2 - R^2 \right)^{\frac{1}{2}} (\Omega) \tag{9.14}$$

where E_O is open circuit e.m.f. per phase at a given value of field current I_f.

I_S is the phase current on short circuit at the *same* value of I_f and R is the resistance of phase winding in ohms.

11. A large supply system is termed an 'Infinite bus-bar'. Its characteristics are

    (i) constant voltage V_S
    (ii) constant frequency f

and   (iii) negligible synchronous impedance Z_s, where

$$Z_s = \sqrt{R^2 + X_s^2}$$

12. A synchronous machine connected to infinite bus-bar cannot alter the system's terminal voltage V_S nor its frequency f.

For a machine to supply or take power from the infinite bus-bar the mechanical prime-mover *must* increase its driving torque or the mechanical load must 'slow' down the machine's rotor.

The variation of excitation, i.e. value of I_f, changes the power factor of the current taken from or supplied to the system.

13. Power input or output to or from the infinite bus-bars is given by

$$P = 3\frac{E_O V_S}{X_s}\sin \alpha \text{ (W)} \tag{9.18}$$

where V_S = bus-bar voltage per phase in volts, and

α = the load angle – i.e. the angle between rotors no-load and operating positions,

α is positive for generator operation and negative for motor operation.

The torque $\qquad T = \dfrac{3}{\omega_R} \cdot \dfrac{E_O V_S}{X_s}\sin \alpha \text{ (Nm)} \tag{9.19}$

14. Characteristics are the curves which:

    (i) relate terminal voltage V_S to load current I_S for a generator operating singly.
    (ii) relate speed of rotation ω_R to the shaft torque T_M for a synchronous motor.

and   (iii) relate T to load angle α for the machine working in parallel with infinite bus-bars as a motor *and* as a generator.

15. Synchronous machines are controlled by

    (i) varying the speed of mechanical prime-mover or the value of the mechanical load

and   (ii) by adjustment of excitation current I_f.

16. Synchronous machines must be synchronised to infinite bus-bars by making their

    (i) potential differences equal
    (ii) frequencies equal
    (iii) phase sequence the same

and   (iv) potential differences in phase.

17. Synchronous motors are not self-starting and therefore must be brought up to near synchronous speed by other means. The methods used are:

    (i) pony motor starting

and   (ii) starting as an induction motor using damper grid windings.

For large machines these methods are combined with starting from reduced voltage supply.

EXERCISES

1. Define in terms of windings used and currents flowing in them the following synchronous machines:

    (i) 'stator fed' type with salient poles on the stator.
    (ii) 'rotor fed' type with salient poles on the rotor.

and   (iii) 'rotor fed' type with cylindrical rotor.

Indicate the range of ratings for each of the above types.

2. (a) What is meant by

    (i) a stationary magnetic field?
    (ii) a rotating magnetic field?

Explain, without proofs, how each is produced. In what type of electrical machine would each of these fields be used?

(b) What is meant by synchronous speed? Give a practical example of the importance of synchronous speed.   (C. & G. 57 Part I 1969)

3. (a) Explain, with the aid of sketches, the principle of operation of a simple two-pole synchronous generator.

(b) Sketch one cycle of alternating e.m.f. and use the sketch to explain the meaning of (i) frequency, (ii) period, (iii) maximum or peak value.

(c) What would be the effect on the generator output of increasing

  (i) the speed of rotation
  (ii) the strength of the magnetic field
  (iii) the number of poles
  (iv) the number of conductors?

(d) What is meant by the r.m.s. value of an alternating voltage? Why is this value used in a.c. work?   (C. & G. 57 Part I 1967)

4. (a) Explain simply the operation of a three-phase 'rotor fed' synchronous machine working as

  (i) a generator
  (ii) a synchronous motor.

(b) A three-phase synchronous machine connected to 50-Hz bus-bars has the following particulars:

Number of rotor poles	= 2
Number of stator turns per phase	= 960
Magnetic flux/pole under full load conditions	= 0·0095 Wb
Winding spread factor	= 0·96
Winding pitch factor	= 0·985
Full load current in each phase	= 350 A
Full load power factor	= 0·8 lagging

Calculate for full load conditions,

  (i) the rotor speed
  (ii) the e.m.f. of rotation generated in each phase winding
  (iii) the power converted by the machine
  (iv) the torque at which electro-mechanical power conversion takes place.

Answers: (i) 3000 rev/min (ii) 1910 V (iii) 1·61 MW (iv) 5120 Nm.

5. (a) Describe a three-phase turbo-generator indicating clearly the types of windings employed in its construction.

(b) A three-phase six-pole star-connected synchronous machine runs as a motor from 11 000-V 50-Hz bus-bars. When the current drawn from the supply is 100 A at a leading power factor of 0·9 and the back e.m.f. is equal to the bus-bar p.d., calculate the speed, power and torque developed by the motor. Neglect all losses.

Answers: 1000 rev/min, 1·88 MW, 1800 Nm.

6. (a) The diagram below shows an equivalent circuit of one phase of a three-phase synchronous machine.

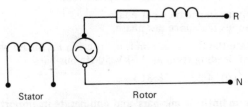

Stator   Rotor

Fig. 9.33   Diagram for question 6

Explain what each element shown represents in an actual machine.
   Label the diagram and show all currents and voltages in it when the machine is working as

    (i) a generator
    (ii) a synchronous motor.

Write down the voltage equation for each mode of operation.

   (b) If the machine is used as a generator supplying its own load, sketch the curves of terminal voltage against load current for different values of load power factor.

   (c) What would be the speed/torque characteristic of the machine when used as a synchronous motor?

7. An 11-kV, 60-MVA, two-pole three-phase star-connected generator has a resistance of 0·04 ohm per phase and a leakage reactance of 0·6 ohm per phase.
   The Open Circuit curve is given below:

Field current (A)	0	300	600	900	1200	1500	1800
Line e.m.f. (kV)	0	3·5	7	10·5	13·4	15·2	16·2

If the field current equivalent to the effect of armature reaction at full load is 300 A, find

    (i) the line e.m.f.
    (ii) the 'equivalent' field current which would produce this e.m.f.
    (iii) the actual current flowing in the machine's field winding
    (iv) the load angle for a full load current at 0·8 power factor lagging.

Answers: (1) 13·9 kV   (ii) 1270 A   (iii) 1510 A   (iv) 19·45°.

8. (a) Explain the term 'synchronous impedance' of a synchronous machine and show how it can be estimated from Open and Short Circuit tests.

   (b) A three-phase generator has a rated output of 1·5 MVA at a terminal voltage of 6·6 kV. The stator winding has a resistance of 1 Ω/phase and a synchronous reactance of 8 Ω/phase.

Calculate the per unit voltage regulation at a power factor of

    (i) unity
    (ii) 0·8 lagging
and   (iii) 0·9 leading.

Sketch a phasor diagram for each case.

Answers: (i) +0·0685   (ii) +0·209   (iii) −0·053.

9. A three-phase, 11-kV 'rotor fed' synchronous motor has a star-connected stator winding with a resistance per phase of 0·3 Ω.

When the machine is tested as a generator with its terminals short-circuited, the full load current of 1400 A is produced by the field current of 600 A. The same excitation gives 10·3 kV between the lines on open circuit.

Find (i) the synchronous impedance per phase
        (ii) the synchronous reactance per phase,

and use these to estimate the fictitious e.m.f. of rotation E_O/phase when the motor draws 24 MW at 0·9 p.f. leading from an 11-kV 50-Hz bus-bars.

Answers: (i) 4·24 Ω (ii) 4·23 Ω 8·31 kV.

10. (a) Explain the term 'infinite bus-bars' and enumerate its important characteristics

   (b) A three-phase star-connected synchronous motor is connected across a 400 V supply. The stator winding has a synchronous reactance of 2 Ω/phase and negligible resistance. If the input power remains constant at 30 kW, find the e.m.f. of rotation per phase for the following power factors:

   (i) unity
   (ii) 0·8 lagging
   (iii) 0·8 leading.

Sketch a phasor diagram for one phase of the motor in each case.

Answers: (i) 246 V   (ii) 185 V   (iii) 286 V.

11. A three-phase 4-pole 6·6-kV, 8-MVA, 50-Hz 'rotor fed', star-connected synchronous machine has a synchronous reactance of 2 Ω/phase and negligible resistance. The excitation is adjusted to give and e.m.f. of rotation E_O 15% higher than the bus-bar p.d. of 6·6 kV to which it is connected.

Find (i) the power factor
        (ii) the load angle in electrical and mechanical degrees
        (iii) the power
and   (iv) the torque

when converting full load MVA as

   (a) a generator
and   (b) a motor.

Neglect rotational losses.

Answers:
Generator: (i) 0·969 lag (ii) 18·5° elec. 9·25° mech. (iii) 7·75 MW (iv) 49 300 Nr
Motor: (i) 0·969 lead (ii) −18·5° elec. −9·25° mech. (iii) 7·75 MW
              (iv) 49 300 Nm.

12. State why the power factor of a synchronous motor working at a constant load depends on its excitation.

   A three-phase star-connected synchronous motor has an equivalent armature reactance of 5·25 Ω and negligible resistance. The exciting current is adjusted to such a value that the generated phase e.m.f. is 1750 V.

   If the motor is operating from 2200-V three-phase supply at a power factor of 0·9 leading, determine the power input to the machine.   (C. & G. 52 Part II 1967)

Answer: 910 kW.

13. A three-phase synchronous generator operates on constant voltage, constant frequency bus-bars. Explain the effect of variation of (a) excitation, (b) steam supply on the power output, power factor, armature current (current flowing in three-phase winding), and load angle of the machine.

An 11-kV, three-phase, two-pole star-connected generator delivers 30 MVA at unity power factor when connected to 'infinite bus-bars'. If the synchronous reactance is 3 Ω/phase and the resistance/phase is negligible, find

    (i) the open circuit e.m.f. between its terminals, and

    (ii) load angle α.

Answers: (i) 13·7 kV   (ii) 38·8°.

14. (a) Describe the procedure for synchroning and connecting a three-phase machine to infinite bus-bars. How is the machine's output and power factor adjusted when it operates as

    (i) a generator

    (ii) a motor.

    (b) A three-phase star-connected generator operates on 11-kV, 50-Hz infinite bus-bars. The winding has a synchronous reactance of 10 Ω/phase and negligible resistance.

If the excitation is adjusted to give an open circuit e.m.f. of 13·8 kV, find

    (i) the maximum power output of the machine,

    (ii) the line current,

    (iii) the power factor,

and (iv) the load angle,

    for this condition.

Answers: (i) 26·4 MW   (ii) 1015 A   (iii) 0·78 lead   (iv) 90° elec.

15. Describe one method of starting a large synchronous motor; give a connection diagram and state, in detail, the switching sequence.

                                        (C. & G. 57 Part II 1968)

10

Synchronous-induction motors

The synchronous-induction motor combines the starting characteristics of a wound rotor induction motor with the running characteristics of a synchronous motor. It differs from the synchronous motor which uses damper grids for starting (described in Chapter 9) in that it has an insulated rotor winding which allows resistance to be varied externally during starting. Two basic types are described in this chapter; the first employs a rotor with salient poles, which carry two separate windings, whilst the second uses a cylindrical rotor with one winding only.

The stators in both types are similar and accommodate a three-phase distributed winding.

A typical salient pole rotor and a typical cylindrical rotor are shown in Figs. 10.1 and 10.2 respectively.

The cylindrical rotor machine may be further sub-divided into a motor which has a d.c. exciter permanently in the rotor circuit (auto-synchronous induction motor) and a motor in which the exciter is disconnected from the rotor circuit during starting-up period.

Fig. 10.1   Rotor of synchronous-induction motor showing the salient poles and excitation winding, with three-phase insulated starting winding in the pole faces. (*English-Electric–A.E.I. Machines Ltd.*)

Fig. 10.2 Cylindrical rotor of synchronous-induction motor showing the three-phase distributed winding with overhung d.c. exciter armature. (*English-Electric–A.E.I. Machines Ltd.*)

10.1   Salient pole motor

In this motor three separate windings are employed as follows:

STATOR: a three-phase distributed winding fed with alternating currents.

ROTOR: a concentrated coil winding wound around the stem of each salient pole and carrying direct current; and an insulated three-phase winding distributed in slots cut in the faces of salient poles and carrying alternating currents.

The rotor coil winding operates when the machine runs as a synchronous motor. The rotor phase winding, which is connected to three external resistors via slip-rings, operates when the machine starts up as a wound rotor induction motor.

The circuit connections are shown in Fig. 10.3.

Initially the field switch is opened and the machine starts as a slip-ring induction motor. The rotor phase windings are connected to the three starting resistors and as the speed builds up the resistors are progressively cut out until the slip rings are short-circuited.

In many schemes, but by no means all, the rotor coil winding during this period is connected to a discharge resistor to limit the induced voltages due to stator's rotating magnetic field. When the rotor reaches its maximum subsynchronous speed, the field switch is closed and the exciter injects direct current into the coil winding, thus producing the magnetic poles on the rotor. These exert the pull on the rotating poles and when the force is sufficiently great and the speed difference small enough the rotor pulls into synchronism. The machine begins to work as a synchronous motor. The rotor starting windings remain shorted during normal operation and serve in the same way as grids to damp down rotor oscillations at load changes and other external disturbances.

The machine has an advantage over the cylindrical rotor type in that each of the two rotor windings can be independently designed.

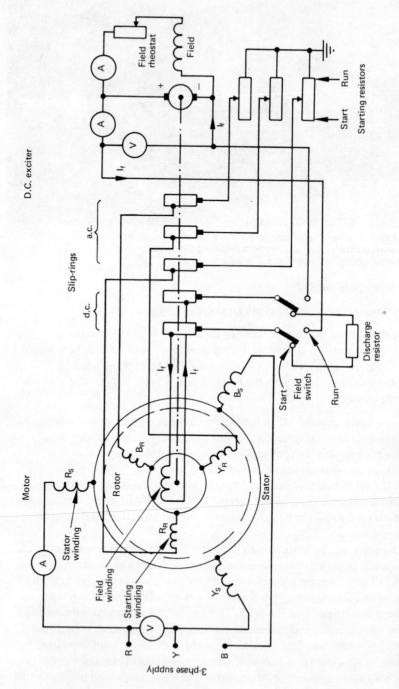

Fig. 10.3 Circuit diagram of a salient pole synchronous-induction motor

10.2   Cylindrical rotor motor

In this machine two windings only are employed; one on the stator and the other on the rotor.

The stator winding is identical with that of a salient pole motor. The rotor winding is an insulated three-phase type which serves two purposes:

      (i) for starting as a wound induction motor

and   (ii) for d.c. excitation during synchronous motor operation.

At synchronous speed the rotor winding is fed with direct current, which is obtained from a d.c. exciter (d.c. generator) or from an a.c. exciter (a.c. generator) together with solid state rectifiers.

In smaller motors up to approximately 1200 kW in rating, d.c. exciters are permanently connected to the rotor winding. The resulting machine is termed an auto-synchronous-induction motor.

The rotor of a d.c. or an a.c. exciter is usually mounted on the main motor shaft. When an a.c. exciter is employed the rectifier bridge is also accommodated in a separate steel hub fixed to the shaft.

In Fig. 10.4 the circuit diagram of an auto-synchronous-induction motor is shown.

The d.c. exciter used in this scheme is permanently connected in one phase of the rotor winding. Initially high voltage at supply frequency is induced in the rotor. The voltage drives an alternating current through all the starting resistors via three connections, one of which has a slightly larger resistance because it includes the commutator winding of the exciter. The voltage developed across the exciter's armature is applied to the exciter's shunt field, but being small and alternating at supply frequency, the exciter has no time to develop its own d.c. output. As the speed of the rotor increases, when the starting resistors are being progressively reduced, the frequency of the rotor currents decreases, until at full subsynchronous speed it is about 0·5 Hz.

As full synchronous speed is approached the shunt wound exciter becomes self-exciting injecting an increasing direct voltage into the main motor's phase winding and tending to create fixed magnetic poles. When their strength is sufficient to lock with the stator's rotating poles, the rotor accelerates to synchronous speed and the machine changes from induction to synchronous motor operation.

The exciter's shunt field rheostat however must be adjusted to give sufficiently high d.c. e.m.f. of rotation to achieve synchronisation. Once the rheostat is preset at this value the synchronising is achieved, automatically – hence the word 'auto' is added to the motors designation.

Every cylindrical rotor with one three-phase winding for both starting and excitation duty must satisfy the conflicting requirements of *low exciting current and low starting voltage* between the slip-rings. These requirements lead to a compromise in choosing the number of turns of the rotor winding. (Small number of turns for low starting e.m.f. and a large number of turns for low exciting current.)

The limit at which an acceptable choice can be made, generally occurs at motor's rating of approximately 1200 kW, when the slip-ring starting voltage is about 2000 V.

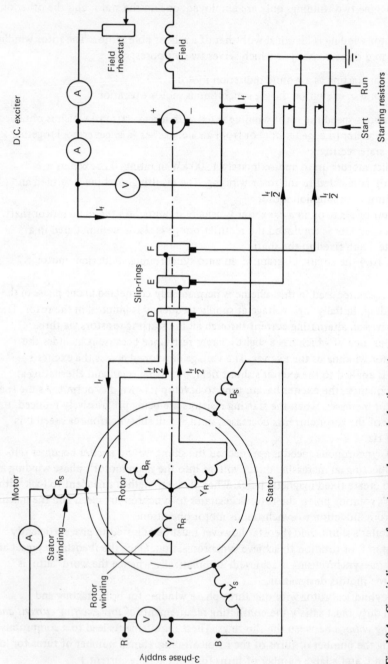

Fig. 10.4 Circuit diagram of an auto-synchronous-induction motor

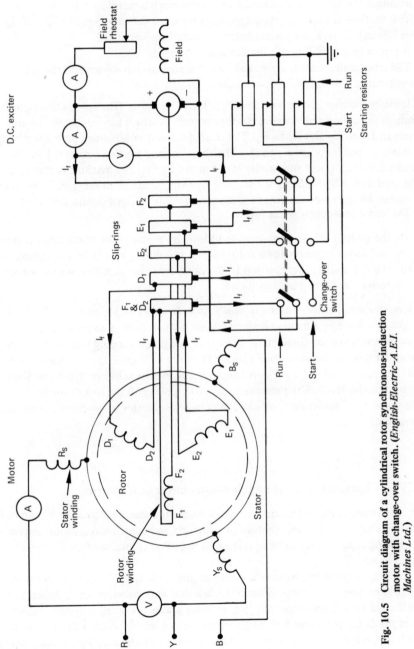

Fig. 10.5 Circuit diagram of a cylindrical rotor synchronous-induction motor with change-over switch. (*English-Electric–A.E.I. Machines Ltd.*)

For larger machines than that it is necessary to connect the rotor winding in delta at starting and then to reconnect it in star for synchronous motor operation in order to maintain the balance between the two requirements mentioned above.

This method reduces the starting voltage between the slip-rings to $1/\sqrt{3}$ of the value it would have if the permanent star connection were used.

A motor incorporating this feature is shown in Fig. 10.5.

The scheme illustrated requires an external change-over switch and the rotor with five slip-rings. The operation is as follows:

Initially the change-over switch is in 'START' position. The rotor's three-phase windings are connected in delta, the two corners of which are formed by joining terminals D_1 to E_2 and E_1 to F_2. The third delta corner is formed permanently by an internal connection of F_1 to D_2. The three corners are also connected to the starting resistors, whilst the exciter is not in the circuit. The machine is started as an induction motor by reducing the starting resistors. When a full sub-synchronous speed is reached by the rotor the switch is changed quickly to position marked 'RUN'.

The new connections are as follows:

(i) the phase $F_1 F_2$ is short-circuited through the star point of starting resistors and serves to damp down rotor oscillations during synchronous running,

(ii) the d.c. exciter is connected to the other two phases, which are reconnected in series, i.e. D_2 is joined to E_2.

The exciter supplies the current I_f which flows from D_1 to D_2 and then from E_2 to E_1, thus producing permanent magnetic poles on the rotor. This condition is equivalent to an instant when the three-phase current in one phase winding is zero and 0·866 of its maximum value in the other two phases. The m.m.f. diagram therefore is the same as that illustrated in Chapter 6 Fig. 6.16(c), except that its position relative to the rotor's surface remains fixed. The permanent magnetic poles pull the rotor into synchronism with the stator's equivalent poles and the machine changes to synchronous mode of operation.

10.3   Characteristics of synchronous-induction motors

The speed/torque charactersitics for synchronous-induction motors are shown in Fig. 10.6 and consist of the combination of induction and synchronous motor curves. The change-over from one mode of operation to the other is indicated by a series of dotted lines.

Starting from the left hand side of the diagram, the first change-over point shown is that at no load. The speed difference is smallest and therefore small direct current is sufficient to pull the rotor into step with the stator's rotating magnetic field. When the motor is started under full load, (second dotted line from the left), the speed at which it operates as an induction machine is lower (the slip may be around 0·06) and therefore the 'pull-in' torque required is very much larger. In this case the exciter's shunt field rheostat must be set to a lower value to allow a higher injection of direct current into the the rotor to produce strong magnetic poles on it.

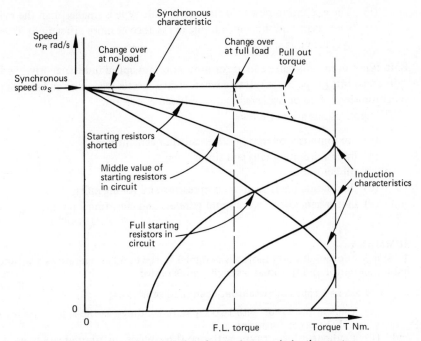

Fig. 10.6 Speed/torque characteristics of a synchronous-induction motor

The third dotted line indicates the point at which the load torque exceeds the maximum synchronous torque the motor is capable of providing and the machine changes back to induction motor operation.

It must be noted that in the latter mode of operation it is possible to drive larger load than under synchronous operation. Unless the machine is designed for continuous operation in this manner, it should not be used in this way for long periods.

10.4 Comparison and application

The majority of synchronous-induction motors are manufactured in ratings ranging from: 50 kW to 2500 kW for cylindrical rotor types and up to 5000 kW for salient pole types.

They may operate from three-phase supplies of up to 11 kV and at speeds of 2 rev/s and up to 60 rev/s. Because of their complexity they are not usually used for outputs below 50 kW, although smaller machines have been built for specific requirements.

Whilst it is not easy to compare the two types, the following points may be made.

(i) The cylindrical rotor type reaches a higher sub-synchronous speed when starting under load and therefore synchronises more easily under arduous conditions.

(ii) The cylindrical rotor type can be designed to run continuously as an induction motor.

and   (iii) The excitation power of the salient pole type is smaller than the cylindrical
            rotor type and this tends to make the former more efficient at lower
            synchronous speeds.

Both types of motors can be used for most constant-speed industrial drives, with the
cylindrical rotor type more suitable where heavy inertia load exist and starting and
synchronising has to be achieved under full load conditions.

The applications are as follows:

   (a) continuously-operating drives, such as cement mills,
   (b) mine-ventilating fans and sinter fans,
   (c) rubber mills,
   (d) fluctuating loads such as compressors and rolling mills,
and   (e) motor-generators, wood-pulp grinders and centrifugal pumps.

SUMMARY

1. Synchronous-induction motor is designed to start and accelerate as a wound rotor
induction motor and to run as a synchronous motor.

2. There are two types of synchronous-induction motors:

   (i) a salient pole type with two separate windings on the rotor, one for starting
       and the other for running.
and   (ii) a cylindrical rotor type with one three-phase distributed winding on the
           rotor for both starting and running.

3. The speed/torque characteristics for synchronous-induction motors are combinations
of induction motor *and* synchronous motor speed/torque curves.

EXERCISES

1. (a) Explain the difference between a plain synchronous motor and a synchronous-
induction motor.

   (b) Enumerate two forms of synchronous-induction motor most commonly used.
Describe their constructional differences and indicate where each may be used in
practice and why.

   (c) 'Large round rotor synchronous-induction motors have two different rotor
winding connections for starting and for running'. Give reasons for the above statement
and suggest the connections employed.

2. Draw a circuit diagram for a three-phase auto-synchronous motor and explain the
method employed in starting this machine. What are the advantages of this type of
motor when compared with (a) a plain synchronous motor and (b) an induction
motor?

   A factory has an average three-phase induction motor load of 1200 kVA at 0·75
power factor. An auto-synchronous motor is employed to improve the total power
factor and at the same time supply a mechanical output of 200 h.p. (149·2kW). Calculate
the kVA input to the auto-synchronous motor and its power factor if the overall average
power factor of the factory is raised to 0·9 lagging. Neglect the losses in the auto-
synchronous motor.

Answers: 320 kVA,   0·465 leading.

3. Sketch speed/torque characteristics for a synchronous-induction motor, assuming
three taps on the starting resistors.

Indicate on it:

     (i) synchronous running charactersitic
    (ii) induction running characteristic
  (iii) induction starting charactersitic
and (iv) pull-out torque.

4. (a) The windings of rotating electrical machines may be grouped under three main headings, namely concentrated, phase and commutator. State which winding is used on the stator and rotor of each of the following types of machine and whether they carry a.c. or d.c. currents.

  (i) Synchronous machine
 (ii) Induction motor

(b) The stator of a machine has a three-phase winding, equivalent to two poles mounted on it, and supplied from a three-phase 50-Hz supply. Show how such a winding produces a rotating magnetic field. Calculate the speed of the rotor when placed inside the stator if the former carries,

  (i) A concentrated winding supplied with d.c. current,
 (ii) a squirrel cage winding if the slip is equal to 0·07.

Answers: (i) 3000 rev/min. (ii) 2790 rev/min.

11 Small electric motors

The development of electrical distribution network throughout the world, but especially in industrially developed countries enables cheap, clean and flexible power to be available to everyone.

The electrical power however must usually be reconverted to mechanical power (energy of motion) to perform the innumerable tasks required. The device for doing just this is a small electric motor.

This chapter is devoted to electro-mechanical convertors which are included in products such as power tools, business machines, control equipment, heating and ventilating equipment, recording and sound reproduction equipment but above all in many domestic appliances. No home nowadays is complete without at least three small motors driving a vacuum cleaner, a washing machine and a record player.

Whilst there is a large variety of these machines, three groups cover nearly all those employed in the areas enumerated above. These are:

      (i) Universal a.c./d.c. motors
      (ii) Single phase induction motors
and   *(iii) Small synchronous motors.*

All are built to work from a single phase a.c. supply. In similar industrial applications where a three-phase supply is available, generally small three-phase motors are used. Their theory is the same as for large machines and is given in the preceding chapters.

11.1 Universal motor – definition and construction

The universal motor is a 'direct current' series excited machine with a stator built up of laminations, which can operate from either a.c. or d.c. supply.

Two forms of stator construction are used:

      (i) a salient pole stator which has only one field winding
and   (ii) a cylindrical stator with two coil windings distributed in its slots; the first serving as a field winding, whilst the second, displaced 90 degrees from the first, as a compensating winding to neutralise armature reaction.

The first type operates satisfactorily for ratings below approximately 250 W ($\frac{1}{4}$ h.p.) the second above this value. The secioned view in Fig. 11.1 illustrates the construction of the first type.

Fig. 11.1 Sectioned view of a universal a.c./d.c. motor rated 130 W ($\frac{1}{5}$ h.p.) at 8500 rev/min. (*English-Electric–A.E.I. Machines Ltd.*)

11.2 Universal motor – simple explanation of operation

The production of a unidirectional torque is easily seen from a motor's circuit diagrams. In Fig. 11.2(a) the circuit is marked for the positive half cycle of an alternating current flow, whereas in Fig. 11.2(b) the negative half cycle is shown. The time graphs of the two half cycles are given showing current and the two m.m.f.s

Both m.m.f.s are produced by the *SAME* current, and therefore reverse simultaneously with the current reversals. The torque being due to the interaction of F_S, and F_R acts always in the same direction, (anticlockwise in Fig. 11.2), although its value varies from zero to maximum and back to zero. The result is a pulsating torque of *always* positive value.

It might be mentioned that at first glance the shunt connected d.c. machine should also work from an a.c. supply, since the voltage reverses simultaneously across both its windings. The currents through the two parallel circuits however *do not* reverse at the same time, because the shunt field reactance is very much larger than that of the commutator winding. The phase angle between the two currents is in fact very large and the torque produced is minimal. Its value may not even be sufficient to run the motor at no load.

11.3 Universal motor – simple phasor diagram and characteristics

A simplified circuit diagram of a universal motor connected to an a.c. supply is shown in Fig. 11.3(a). The resistances of a field coil and a commutator winding as well as the reactance of the latter are all neglected since their values are small compared with

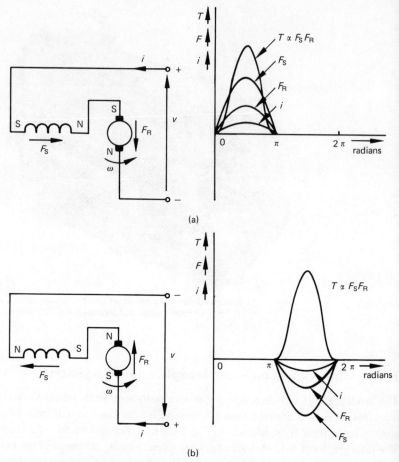

Fig. 11.2 Torque in a universal motor

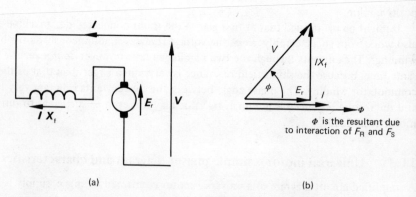

Fig. 11.3 Circuit and phasor diagram for a universal motor working on
a.c. supply

the magnetising reactance of the field winding and the value of an e.m.f. of rotation E_r. Applying Kirchhoff's voltage law to the circuit,

$$V = E_r + IX_f \qquad (11.1)$$

The above equation is solved graphically by means of a phasor diagram in Fig. 11.3(b).

E_r is drawn in phase with, and IX_f 90° ahead of the current I. This is so because the current produces the flux which when 'cut' by the conductors, induces in them an e.m.f. E_r, hence current, its flux and the e.m.f. are all in phase with each other.

Neglecting all losses and applying fundamental equation $T\omega = E_r I$, the torque developed at the shaft is given by

$$T_{\text{a.c.}} = \frac{E_r I}{\omega} = \frac{(V \cos \phi) I}{\omega} = \frac{VI \cos \phi}{\omega} \text{ (Nm)}$$

i.e.

$$T_{\text{a.c.}} = \frac{VI \cos \phi}{\omega} \text{ (Nm)} \qquad (11.2)$$

where $VI \cos \phi$ = electrical input in W.

When the motor is operated from a d.c. supply and again neglecting all losses, then

$$T_{\text{d.c.}} = \frac{VI}{\omega} \text{ (Nw)} \qquad (11.3)$$

thus at the same values of V, I and speed ω the ratio of:

$$\frac{T_{\text{a.c.}}}{T_{\text{d.c.}}} = \frac{VI \cos \phi}{VI} = \cos \phi = \text{power factor} \qquad (11.4)$$

Equation (11.3) shows that series motor develops less torque when operating from an a.c. supply than when working from an equivalent d.c. system.

Typical speed/torque characteristics are given in Fig. 11.4. These are the same in shape as for an ordinary series excited d.c. motor. The a.c. characteristic lies below the d.c. curve.

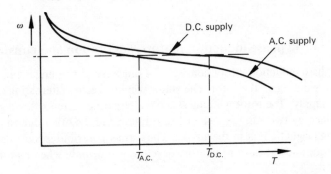

Fig. 11.4 Typical speed/torque characteristics of a universal motor

11.4   Universal motor – control

Although universal motors can operate from either a.c. or d.c. supplies, nevertheless they are designed and usually operate from the former.

They are started by direct connection to the supply, and in case of larger units through a 'direct on line' contactor.

Their speed can be adjusted if the field coil is provided with tappings, so that its number of turns can be varied. Circuit diagram of variable speed motor is shown in Fig. 11.5(a).

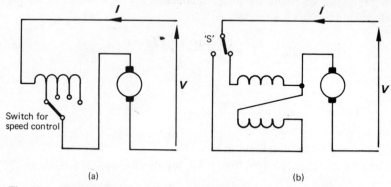

(a)   (b)

Fig. 11.5   (a) Speed variation (b) Reversing. Control of a universal motor

The direction of rotation of a motor can be controlled by changing over the connections to one of its windings, *provided* that the brushes are fixed along the flux neutral axis. If they are mounted permanently slightly backwards to improve the commutation the motor is not reversible.

Alternatively a motor can be provided with two field windings mounted on the stator; one for each direction of rotation. The selection is then made by means of a switch 'S' as shown in Fig. 11.5(b).

Since the speed of these motors is not limited by the supply frequency and may be as high as 20 000 rev/min., they are suitable for such application as high-speed hand tools. They are also employed in vacuum cleaners, sewing machines and many other areas where speed control and high values of speed are necessary.

11.5   Single phase induction motors – definition and construction

A single phase induction motor consists of a single phase winding mounted on a stator and a cage winding on the rotor. The stator winding carries alternating current drawn from the supply. The rotor winding also has alternating currents flowing in it, which are produced by two e.m.f.'s; one due to rotation and the other due to pulsation of an alternating magnetic flux in the air-gap. The former contributes towards production of the torque and the motor therefore develops torque only when its rotor is revolving.

At starting, extra provision must be made to initiate rotation, and according to which method is employed the machine is classified as:

    (i) induction motor with split-phase start
    (ii) induction motor with capacitor start
    (iii) induction motor with capacitor start and run
    (iv) induction motor with permanent-split capacitor
    (v) induction motor with shaded poles.

The sectional view of a capacitor start machine illustrated in Fig. 11.6 shows constructional features of a typical small induction motor.

Fig. 11.6   Sectional view of a single phase capacitor motor: output range
125-375 W ($\frac{1}{6}$-$\frac{1}{2}$h.p.) with 2, 4 and 6 poles. *(English-Electric-A.E.I.*
Machines Ltd.)

11.6   Induction motor – operation

The cross-section of a two-pole machine is shown in Fig. 11.7(a). Initially the rotor is stationary and the stator winding is supplied from an a.c. source.

    At an instant marked XX$'$ in Fig. 11.7(b) the stator current $i_s = I_m$ produces an alternating m.m.f. F_S which can be represented by a space phasor. (See Chapter 6.)

    The cage winding conductors maybe looked upon as being divided into two groups. Those in group YY$'$ form coils, the planes of which are perpendicular to, and those in group XX$'$ in line with the stator m.m.f. F_S. (In actual machines the dividing line between the two groups is not at all clear cut.)

    The m.m.f. F_S produces sinusoidally distributed alternating flux in the air-gap, which links the coils YY$'$ and induces in them an alternating e.m.f. The e.m.f. in turn

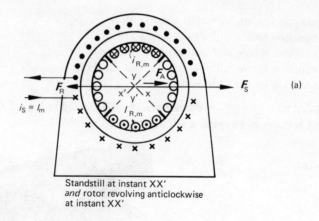

Standstill at instant XX'
and rotor revolving anticlockwise
at instant XX'

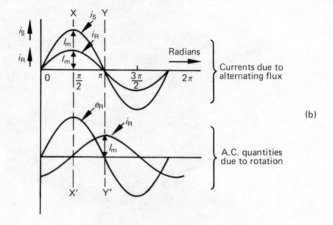

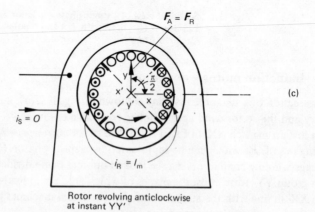

Rotor revolving anticlockwise
at instant YY'

Fig. 11.7 Appertaining to single phase induction motors

produces the current $i_{R,m}$ shown in Fig. 11.7(a). By Lenz's law $i_{R,m}$ gives rise to the rotor m.m.f. F_R, which opposes F_S. The difference $F_A = F_S - F_R$ is the resultant air-gap m.m.f. F_A.

The machine behaves as a transformer with short-circuited secondary (rotor). No lateral force is generated and the rotor does not revolve.

The same machine is now considered but with the rotor revolving. The rotor is turned first rapidly by a mechanical effort in an anticlockwise direction.

The instant illustrated is again that marked by XX' in Fig. 11.7(b), that is when the stator current is a maximum. The distribution of the currents in the YY' windings is exactly the same as shown in Fig. 11.7(a), but in addition the e.m.f. of rotation e_r is generated in conductors XX' due to the air-gap flux being cut by the rotor conductors. The direction of the e.m.f. deduced by Fleming's R.H. rule is as follows:

section X into the plane of the paper
section X' out of the plane of the paper.

In an induction machine the ratio of magnetising reactance X_R to resistance R_R of the rotor winding is always large, for the operation of the machine depends on its value being much greater than unity, i.e. $X_R/R_R \gg 1$.

For simplicity the resistance R_R is neglected here and therefore the rotor current i_R lags behind the rotational e.m.f. e_r by $\pi/2$ radians (Fig. 11.7(b)).

Thus at an instant XX' the current i_R in conductors XX' is zero, although the e.m.f. e_r is a maximum.

The air-gap m.m.f. is therefore the same as at standstill, i.e., F_A directed from left to right (Fig. 11.7(a)).

The machine at an instant marked by the line YY' in Fig. 11.7(b) is illustrated in Fig. 11.7(c).

At this instant the stator current, the rotor current in conductors YY' due to alternating flux only and the rotor e.m.f. of rotation are all zero.

The rotor current i_R in conductors XX' is however a maximum and produces an m.m.f. vertically upwards. This is the only m.m.f. present and therefore constitutes an air-gap magneto-motive force F_A.

Hence the airgap m.m.f. F_A maybe considered as having rotated through $\pi/2$ radians in an anticlockwise direction.

These two instants are sufficient to show that a *rotating m.m.f.*, and therefore magnetic field is established in a single phase induction motor *when its rotor is revolving*.

The effect is similar to that produced by a two-phase winding described in Chapter 6, except that the magnitude of the field is not constant throughout each revolution. In fact its value depends on the speed of the rotor itself. The rotor cage performs the duty of a second phase winding and makes the creation of a rotating m.m.f. possible.

The speed of the magnetic field can be deduced as for a three-phase machine, i.e.,

$$\omega_S = \frac{2\pi f}{p} \text{(rad/s)} = \frac{f}{p} \text{(rev/s)} \tag{11.5}$$

where f = frequency of the supply in Hz
and p = number of equivalent pole pairs on the stator.

Just as in case of a three-phase machine (Chapter 8) the rotating air-gap flux reacts with the currents in a cage winding and the rotor picks up the speed until forward torque is balanced by the opposing torque. This occurs at a speed ω_R which is always lower than that of the magnetic field. The slip for this motor is defined in the same way as for the three-phase machine, i.e.,

$$s = \frac{\omega_S - \omega_R}{\omega_S} = \frac{n_S - n_R}{n_S} \tag{11.6}$$

It is to be noted that the values of s for a single phase motor are greater than in the case of an equivalent three-phase machine because of the variable magnitude of the rotating magnetic flux.

If the rotor is turned mechanically in a clockwise direction then its conductors cut the stator flux in opposite direction to that of the case just described and the rotating field is produced exactly as explained already but it revolves in a clockwise direction. The motor therefore runs in that direcion.

Thus the original impulse on the rotor determines the direction of rotation of a single phase induction motor.

In such a motor it is essential for the ratio of X_R/R_R to be high in order that the phase angle between the stator and rotor currents is as near $\pi/2$ radians as possible.

X_R is the magnetising reactance of the rotor winding and must not be confused with the leakage reactance of that winding. In order to obtain high value of X_R the air-gap in induction motors is made as small as permitted by mechanical consideration. This gives high permeability of the magnetic path in the direction of the m.m.f. F_R.

The behaviour of the machine with a single stator winding may be summarised as follows:

    (i) At standstill, only an alternating flux exists in the air-gap and no torque is exerted on the rotor.

    (ii) Once the rotor is revolving the rotating magnetic field is set up in the air-gap which is due to:

        (a) stator m.m.f. F_S.

and   (b) rotor m.m.f. F_R produced by the rotation through the stator magnetic flux. Hence the torque is exerted on the rotor and the motor continues to run.

    (iii) The rotating field is of variable magnitude throughout each revolution.

    (iv) The rotating magnetic field can be obtained in either direction depending on the original turning effort given to the rotor.

11.7   Induction motor – speed/torque characteristic

A speed/torque characteristic can be obtained experimentally by running the motor from a constant voltage supply, loading it with a brake and measuring speed and torque at different values of the mechanical load.

When the values are plotted then the resulting curve is similar in shape to that for a three-phase machine and is shown in Fig. 11.8.

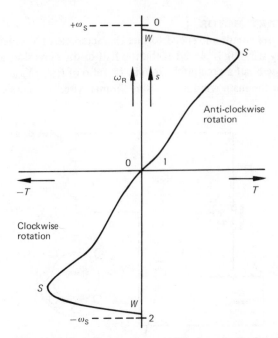

Fig. 11.8 Speed/torque characteristic for a single phase induction motor

The curve indicates clearly that

   (i) the torque is zero at standstill ($s = 1$)
  (ii) the motor develops torque in either direction of rotation
and (iii) the operating portion of the characteristic marked WS is nearly horizontal,
         i.e., the machine is approximately a constant speed motor.

11.8   Induction motor – starting

As shown already, a single phase motor has no torque at standstill and therefore,
means must be provided for starting it. Mechanical methods are impractical and so
the motor is started by temporarily converting it into a two-phase machine by
addition of an auxiliary stator winding. An auxiliary winding, if supplied with a
current I_{AW} which is out of phase with the main current I_{MW} would produce in
conjunction with the main or running winding a *ROTATING MAGNETIC FIELD*,
which in turn generates an initial torque (see Chapter 6).

   Clearly the same single phase supply must be used for both windings. Hence to
produce phase difference between I_{MW} and I_{AW}, the impedances of the two windings
must be different. Several methods exist for doing this and these give rise to different
motor designations. Each arrangement however is inferior to that of a pure two-phase
motor, because the time displacement between the two currents cannot be made
equal to $\pi/2$ radians. Therefore an amplitude and an angular velocity of the rotating
field is never constant. Nevertheless the results are adequate for starting purposes
and the most common arrangements found in use are discussed below:

SPLIT-PHASE START MOTOR

Figure 11.9 shows the circuit and speed/torque characteristic of a typical split-phase motor. An auxiliary winding is placed at the top half of the same slots as the main winding but displaced half a pole pitch from it. Its ratio of R_{AW}/X_{AW} is made greater than that of the main winding by using thinner wire, i.e., increasing its resistance.

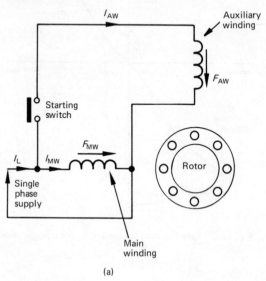

(a)

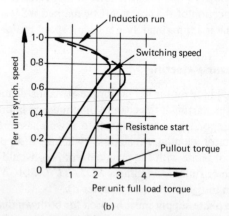

(b)

Fig. 11.9 (a) Circuit (b) Speed/torque characteristics. Split-phase motor

This winding, rated for a short time, is disconnected from the supply at approximately 75 per cent of synchronous speed by a centrifugally operated switch. The speed/torque characteristics in Fig. 11.9(b) show that:

(a) an induction run gives smaller speed variation *without* the auxiliary winding above approximately 85 per cent of synchronous speed,

(b) a starting torque = 1·5 times full load torque, and

(c) a pull out torque = 2·3 to 2·5 times full load torque.

The starting current however is high, 7 to 8 times full load value.

These motors are well suited for loads where the inertia is low and frequency of starts limited, for example, air-conditioning fans, blowers, centrifugal pumps, small drills, lathes, food mixers, office machinery, dairy machinery, floor polishers, etc.

CAPACITOR-START, CAPACITOR START/CAPACITOR RUN AND PERMANENT SPLIT CAPACITOR MOTORS

Apart from the capacitor-start motor, the other two types employ an auxiliary winding which is rated for continuous operation.

The circuit diagrams in Figs. 11.10(a), (b) and (c) show an external capacitor connected in series with an auxiliary winding in order that the phase angle between I_{MW} and I_{AW} may be as near $\pi/2$ radians as possible.

The value of starting capacitor must be large to give high initial torque, whereas smaller capacitor is required for good performance near the synchronous speed. This is so, because initially the main winding takes a very large starting current I_{MW}, and so the auxiliary winding circuit must have low capacitive reactance ($X = 1/2\pi f C_S$) to give comparable value of I_{AW}. During normal working, however, the current in the main winding diminishes and to restore the current balance in the two windings, the capacitive reactance must be increased, which necessitates a smaller value of C_R,

$$\text{since } X = \frac{1}{2\pi f C_R}.$$

Therefore a capacitor start/capacitor run motor employs two separate capacitors to obtain the best performance at starting and also whilst operating under loaded conditions. Thus its operation approximates to that of a balanced two-phase machine.

Permanent split capacitor motor does not employ a starting switch, but at the expense of low starting torque.

The characteristics shown in Fig. 11.10(d) indicate clearly salient features of each type of motor. Typical values of starting torque, pull-out torque and the quality of an operating portion of the curves may be readily estimated from them.

It must be noted that in addition to obtaining a high starting torque with the use of a large capacitor the initial line current drawn by the motor is reduced to 4–4·5 times its full load value, due to a large phase angle between the two winding currents.

Typical applications are as follows:

Capacitor start motors; loads of higher inertia requiring frequent starts such as pumping equipment, refrigeration and air compressors.

Capacitor start/capacitor run motors: as above but where the maximum pull out torque and efficiency required are higher, combined with quieter running.

Permanent-split motors: widely used to drive ventilating fans; capable of speed variation by adjusting the voltage applied to the auxiliary winding through a small single phase auto-transformer.

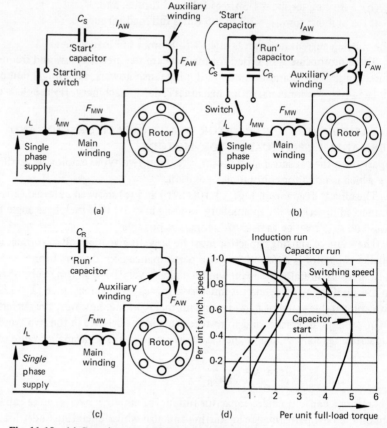

Fig. 11.10   (a) Capacitor start (b) Capacitor start/capacitor run
(c) Permanent split capacitor (d) Speed/torque characteristics.
Induction motors employing capacitors

SHADED POLE MOTORS

The methods so far discussed depended on converting a single phase motor into a temporary two-phase machine by providing an auxiliary winding and a current I_{AW} flowing in it which is out of phase with the current in the main winding.

The object is to obtain two m.m.f.s F_{MW} and F_{AW} displaced both physically and in time phase to each other and therefore capable of combining into a resultant rotating m.m.f. F_A which produces an air-gap flux.

A 'shaded pole' method maybe looked upon as a direct way of splitting the main pulsating flux into two components which are both in time phase and physical displacement to each other. This is achieved by means of a shading ring (a copper band) which surrounds *a part* of a main flux produced by each equivalent pole as shown in Fig. 11.11.

The supply current I_S causes a pulsating flux Φ_S, a proportion of which threads through a shading ring mounted off the centre line of the main flux. The ring behaves as a single short-circuited turn of a transformer's secondary. The e.m.f. induced in it drives an alternating current around it which in turn generates a flux opposing the

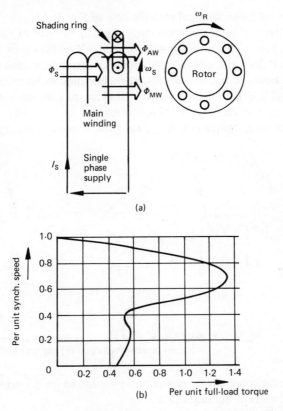

Fig. 11.11 (a) Circuit (b) Speed/torque characteristic. Shaded pole motor

portion of Φ_S linking the ring. The resultant of these two is the flux Φ_{AW} which in an ideal case is 90° out of step with Φ_{MW}.

In this way both the necessary conditions for rotation of magnetic field are obtained and the starting torque is created. The flux revolves always *towards* the shading ring and therefore the motor's direction cannot be reversed.

A typical speed/torque characteristic is given in Fig. 11.11(b), which shows that the starting as well as a maximum stalling torque are low, so also is the starting current.

These motors are particularly suitable for fan drives and are used in domestic and industrial refrigeration, air-conditioning equipment, as exhaust fans, desk fans, hand and hair dryers and electronic equipment cooling fans. They are also employed to drive tape recorders, rotary ironers, etc.

11.9 Single phase synchronous motors

The use of these motors is limited to instruments and electric clocks. The reason is that they can only produce driving torque at absolutely constant speed determined by the frequency of the supply ($\omega_R = 2\pi f/p$) and require in addition to starting devices a permanent-magnet field or a d.c. excitation. In small machines the former method is invariably employed.

The operation can be understood with the help of Fig. 11.12.

During a positive half cycle the stator current produces a pair of magnetic poles as shown in Fig. 11.12(a). The stator 'N'-pole exerts a pull on the rotors 'S'-pole and the latter is pulled in a clockwise direction. When the current falls to zero the stator's pole strength decreases to zero and the rotor's momentum carries its 'S'-pole past the stator's central line. During the negative half cycle the polarity of poles is reversed as shown in Fig. 11.12(b) and the stator S-pole exerts a push on the rotor. The continuous, if pulsating, torque results and the rotor revolves at synchronous speed.

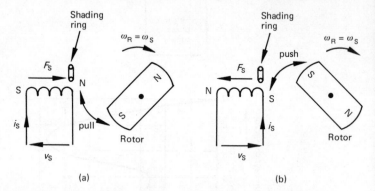

Fig. 11.12 Single phase synchronous motor with a permanent magnet rotor. (a) positive half cycle (b) negative half cycle

It is clear that the rotation can only occur at a speed which is matched exactly to the polarity changes of the stator poles.

Starting torque is provided by means of a shading ring placed in the stator slots and embracing part of the main magnetic flux.

When the motor runs up to near synchronous speed the pull between the stator's and rotor's magnetic poles brings the latter's revolutions up to synchronous speed.

In small domestic electric clocks, recorders, turntables, automatic telegraphy, picture reproduction, etc. unmagnetised rotors are often used. This gives either reluctance or hysteresis type motors, which however produce much smaller values of torque as compared with normal synchronous machine described here.

SUMMARY

1. Widespread availability of electric power necessitates demand for small electro-mechanical convertors to work off a single phase supply. These fall within three main groups:

     (i) a.c./d.c. universal motors
     (ii) single phase induction motors
and   (iii) small synchronous motors.

2. Universal motor is a 'direct current' series connected machine which works from either a.c. or d.c. supply.

3. Expression for torque of a universal motor working from a.c. supply is

$$T_{a.c.} = \frac{VI \cos \phi}{\omega} \text{ (Nm)}$$

(11.2)

and working from d.c. supply is

$$T_{d.c.} = \frac{VI}{\omega} \text{ (Nm)} \tag{11.3}$$

hence the ratio of

$$\frac{T_{a.c.}}{T_{d.c}} = \cos\phi = \text{power factor} \tag{11.4}$$

where V = supply voltage in V.
  I = supply current in A.
  ω = motor's speed in rad/s.
and $\cos\phi$ = power factor.

4. Single phase induction motor is defined as having a phase winding on the stator and a cage winding on its rotor both carrying alternating currents.

5. Induction motors are subdivided according to their methods of starting as:

   (i) split-phase start
   (ii) capacitor start
   (iii) capacitor start and run
   (iv) permanent split capacitor
and (v) shaded pole start

6. Summary of behaviour of a single phase induction motor with one stator winding is as follows:

   (i) at standstill – the machine generates a pulsating flux and no torque.
   (ii) in motion – the machine generates rotating magnetic field which is due to
       (a) stator m.m.f. F_S
       (b) rotor m.m.f. F_R produced by rotation through the pulsating stator flux. Hence the torque is produced and the motor continues to run.
   (iii) unlike a two-phase machine, a single winding motor generates magnetic flux which is *not* constant in magnitude.
   (iv) the motor can rotate in either direction depending on the initial impulse given to its rotor.

7. The characteristics of various induction motors consist of speed ω_R plotted against torque and are similar in shape to those of a three-phase induction motor.

8. Single phase synchronous motor normally consists of a single phase type winding on its stator fed with an alternating current and a permanent magnet rotor.

EXERCISES
1. (a) What are the chief differences in construction between a.c./d.c. series motors and d.c. series motors?
   (b) Sketch the speed/torque characteristics of the a.c./d.c. type when used on
(i) a.c. and (ii) d.c.   (C. & G.L.I. ET II, 1961)

2. (a) Explain simply why a universal motor can operate from d.c. as well as a.c. supplies.
   (b) A universal motor takes 3 A from a 240-V, 50-Hz supply whilst driving a mechanical load at 3000 rev/min. If the reactance of the motor windings is 60 Ω and their resistance negligible by comparison, calculate,

    (i) an e.m.f. of rotation,
    (ii) power factor,
    (iii) power consumed by the motor,
    (iv) the torque delivered to the load.

Neglect all losses.

Answers: 158·5 V,   0·66,   475·5 W,   1·51 Nm.

3. Explain why a single phase induction motor (without starting windings) does not develop a torque when the rotor is at rest, but will do so if the rotor is moving in either direction.
    Describe two methods by which the motor may be made self-starting. How may the direction of the rotation be reversed?   (C. & G.L.I. ET II, 1962)

4. Show, by a sequence of diagrams, how a rotating magnetic field is produced by the stator of a single phase capacitor-run motor. Give a diagram of connections and indicate how these must be changed to reverse the direction of rotation.
                                              (C. & G.L.I. ET II, 1969)

5. What is the basic principle by which a single phase induction motor is made to be self-starting?
    Describe one method and state the switchgear and protective devices which should be provided; give a circuit diagram and show how the direction of rotation may be reversed.   (C. & G.L.I. ET II, 1968)

6. (a) Describe construction and a mode of operation of a universal motor giving typical speed-torque curves when operated from:

    (i) a.c. supply   (ii) d.c. supply.

Account for the discrepancies of the two curves and indicate how the direction of rotation of this motor can be reversed.
    (b) Describe an experiment you have performed on a single phase induction motor, giving circuit diagram and typical characteristics obtained.
    (c) Give details of four methods of starting small single phase induction motors, and include typical applications for which these types would be suitable.

Index

Subjects listed in the Contents, (page v), are not included in the Index.